WORKSHOPS IN COMPUTING
Series edited by C. J. van Rijsbergen

*Also in this series*

AI and Cognitive Science '89, Dublin City University, Eire,
14–15 September 1989
A. F. Smeaton and G. McDermott (Eds.)

Specification and Verification of Concurrent Systems, University of
Stirling, Scotland, 6–8 July 1988
C. Rattray (Ed.)

Semantics for Concurrency, Proceedings of the International
BCS-FACS Workshop, Sponsored by Logic for IT (S.E.R.C.), University
of Leicester, UK, 23–25 July 1990
M. Z. Kwiatkowska, M. W. Shields and R. M. Thomas (Eds.)

Functional Programming, Proceedings of the 1989 Glasgow
Workshop, Fraserburgh, Scotland, 21–23 August 1989
K. Davis and J. Hughes (Eds.)

Persistent Object Systems, Proceedings of the Third International
Workshop, Newcastle, Australia, 10–13 January 1989
J. Rosenberg and D. Koch (Eds.)

Z User Workshop, Proceedings of the Fourth Annual Z User Meeting,
Oxford, 15 December 1989
J. E. Nicholls (Ed.)

Formal Methods for Trustworthy Computer Systems (FM89), Halifax,
Canada, 23–27 July 1989
Dan Craigen (Editor) and Karen Summerskill (Assistant Editor)

Security and Persistence, Proceedings of the International Workshop
on Computer Architectures to Support Security and Persistence of
Information, Bremen, West Germany, 8–11 May 1990
John Rosenberg and J. Leslie Keedy (Eds.)

Women into Computing: Selected Papers 1988–1990
Gillian Lovegrove and Barbara Segal (Eds.)

Carroll Morgan and J. C. P. Woodcock (Eds.)

# 3rd Refinement Workshop

Proceedings of the 3rd Refinement Workshop (organised by BCS-FACS, and sponsored by IBM UK Laboratories, Hursley Park and the Programming Research Group, University of Oxford)

9–11 January 1990, Hursley Park

Published in collaboration with the
British Computer Society

Springer-Verlag
London Berlin Heidelberg New York
Paris Tokyo Hong Kong

Carroll Morgan, BSc (Hons), PhD
Oxford University Computing Laboratory
8–11 Keble Road
Oxford, OX1 3QD

J. C. P. Woodcock, BSc (Hons), MSc, PhD,
Oxford University Computing Laboratory
8–11 Keble Road
Oxford, OX1 3QD

ISBN 3-540-19624-2 Springer-Verlag Berlin Heidelberg New York
ISBN 0-387-19624-2 Springer-Verlag New York Berlin Heidelberg

British Library Cataloguing in Publication Data
3rd refinement workshop: organised by the Programming Research Group and IBM UK Laboratories, Hursley Park, 9–11 January 1990. – (Workshops in computing).
  1. Computer systems. Software. Development
  I. Morgan, Carroll 1952–  II. Woodcock, J. C. P. 1956–  III. Oxford University Computing Laboratory *Programming Research Group*  IV. IBM UK Laboratories  V. British Computer Society  V: Series
  005.1
ISBN 3-540-19624-2

Library of Congress Cataloging-in-Publication Data
Refinement Workshop (3rd:1990:IBM UK Laboratories)
3rd Refinement Workshop:organised by the Programming Research Group, Oxford, and IBM UK Laboratories. Hursley Park, 9–11 January 1990/Carroll Morgan and J.C.P. Woodcock.
    p. cm. – (Workshops in computing)
"Published in collaboration with the British Computer Society." Includes index.
ISBN 0-387-19624-2 (U.S.:alk. paper)
   1. Electronic digital computers–Programming–Congresses.
   I. Morgan, Carroll, 1952–  .  II. Woodcock, Jim.   III. Oxford University Computing Laboratory. Programming Research Group.   IV. IBM UK Laboratories.   V. Title.
VI. Title: Third Refinement Workshop.   VII. Series.
QA76.6.R424 1991                                                              90-22992
005.1–dc20                                                                         CIP

Apart from any fair dealing for the purposes of research or private study, or criticism or review, as permitted under the Copyright, Designs and Patents Act 1988, this publication may only be reproduced, stored or transmitted, in any form or by any means, with the prior permission in writing of the publishers, or in the case of reprographic reproduction in accordance with the terms of licences issued by the Copyright Licensing Agency. Enquiries concerning reproduction outside those terms should be sent to the publishers.

© British Computer Society 1991
Printed in Great Britain

The use of registered names, trademarks etc. in this publication does not imply, even in the absence of a specific statement, that such names are exempt from the relevant laws and regulations and therefore free for general use.

Printed and bound by Alden Press Ltd, Osney Mead, Oxford
2128/3830-543210  Printed on acid-free paper

# Foreword

The papers contained in these proceedings were presented at the Third Refinement Workshop which was held at IBM's Hursley Park Laboratory between the 9th and 11th January 1990. In January 1988 Professor John McDermid of the University of York organised a two-day workshop on the Theory and Practice of Refinement which was held at the King's Manor in York. In the foreword to the workshop proceedings John stated that the two main obstacles to the application of formal methods were the lack of an adequate technique for modularising specifications and the limitations of refinement techniques. The purpose of the first workshop was, therefore, to focus attention on refinement and expedite work in the area. Since that first meeting at York two further successful workshops have been held and a fourth is in the process of being organised.

Between the 10th and 12th of January 1989 there followed the Second Refinement Workshop which was held in Milton Keynes and was organised by Ray Weedon of the Open University. The first day featured a tutorial given by John Wordsworth of IBM and Jim Woodcock of the Programming Research Group and the remaining two days were taken up by twelve technical contributions four of which were from the Esprit RAISE project. As the proceedings have not been published I have set out details of the programme below.

**Day 1**
- Tutorials on sequential and concurrent refinement: John Wordsworth and Jim Woodcock.

**Day 2**
- Formal Specification as it relates to large Systems: John McDermid, York University.
- Refinement, Conformance and Inheritance: Elspeth Cusack, British Telecom.
- Refinement, of Secure Systems and Why it is Hard: Jeremy Jacobs, Programming Research Group.
- Refinement Methods for Process Algebras: Nigel Edwards, Hewlett Packard Laboratory.

- Computer Assistance for a Maths-based Development Process: Ib Holm Sørensen, BP Research.
- Refinement Used to Overcome the Infinite – The Implementation of Pattern Matching in ML: Gill Randell, RSRE.
- Refinement of Software Engineering Principles by Underpinning with Formalisms: Nick Bleach, SD Scicon

**Day 3**
- The Functional Interpretation of Records and Variants: Steve Hughes, STC.
- Equational Sequential Specifications and their Refinement: Tony Evans, STC.
- The Implementation Relation in RAISE: Roy Simpson, STC.
- Equational Concurrent Specifications and their Refinement: Chris George, STC.
- Towards a Refinement Tool for CSP: Peter Strain-Clark, The Open University.

During the workshop a discussion took place concerning the need to continue with the series and the need to organise them on a more formal footing. The upshot of the discussion was that the British Computer Society's Formal Aspects of Computing special interest group and the Programming Research Group of Oxford University agreed to run the following workshop in January 1990. An organising committee for the Third Refinement Workshop was set up and IBM UK Laboratories Ltd. kindly offered to sponsor the event by making available their Hursley Park facilities, and providing organisational help and catering. The organising committee were:

| | | |
|---|---|---|
| *Ian Cottam* | Manchester University | Publicity |
| *Tim Denvir* | Praxis | Finance |
| *Mike McMorran* | IBM | Local Arrangements |
| *Carroll Morgan* | PRG Oxford | Technical Programme |
| *Roger Shaw* | Praxis | Chairman |
| *Jim Woodcock* | PRG Oxford | Technical Programme |

The event was very successful, attracting 105 participants from over six countries, and was made the more enjoyable by the hospitality of our hosts. The committee would like to thank everyone involved in organising the workshop for their efforts and in particular Dawn Jeeves of IBM, the conference secretary, who provided us with excellent support both before and during the workshop.

The record of the proceedings contained in this book do not include material from the tutorials presented on the first day of the workshop. I would like to thank Geraint Jones for his tutorial on Refining Functional Programs, Carroll Morgan for his contribution on The Refinement Calculus, and Jim Woodcock for his tutorial on Concurrency.

The Fourth Refinement Workshop is currently being organised and will be held at Wolfson College Cambridge between the 9th and 11th of January 1991. I look forward to meeting you all there and can guarantee an equally stimulating programme.

Roger Shaw<br>
Workshop Chairman

## Reference

J. A. McDermid, Ed., *The Theory and Practice of Refinement – Approaches to the Formal Development of Large Scale Software Systems*, Butterworths, 1989.

# Contents

Introduction ............................................................................. 1

Deriving an Occam Implementation of Action Systems
R. J. R. Back and K. Sere ..................................................... 9

Refinement Algebra Proves Correctness of Compiling
Specifications
C. A. R. Hoare ...................................................................... 33

On the Usability of Logics Which Handle Partial Functions
J. H. Cheng and C. B. Jones ................................................ 51

Designing and Refining Specifications with Modules
J. M. Morris and S. N. Ahmed .............................................. 73

Formal Program Development in Extended ML for the Working
Programmer
D. Sannella .......................................................................... 99

Relations and Refinement in Circuit Design
G. Jones and M. Sheeran ..................................................... 133

Specifying and Refining Concurrent Systems – an Example from
the RAISE Project
C. W. George and R. E. Milne .............................................. 155

Using Refinement to Convince: Lessons Learned from a Case
Study
C. T. Sennett ....................................................................... 171

Author Index ........................................................................ 199

# Introduction

Refinement is the process of turning specifications into implementations. Of particular interest these days are refinement methods in which many small steps take a specification ever closer to a correct implementation of it.

It is obvious that all users of computers should be interested in refinement, for the problem of correctness is universal. And indeed many users are — they are the ones who wish to gain control over the process of making computer systems that work. They may be clients, vendors, programmers, or even academics.

But alas this volume does not present a single refinement method that is guaranteed to work for all problems. Indeed, that will never be possible. Instead the reader will find here a collection of methods that individually apply to certain broad situations. Also to be found is work on the basis for refinement itself.

That there are many different refinement methods does not represent disagreement, only that different refinement methods suit different applications. In this collection there are four: sequential imperative programs, functional programs, concurrent or distributed programs, and digital hardware circuits.

Sequential imperative programs are those written in conventional programming languages like FORTRAN, Pascal, COBOL, or C. The theory necessary for their correctness has been available for twenty years — but its use has evolved, and it was not always called refinement. At first, the style was to write a program then prove it correct; later, programs and their correctness proofs were developed together; only more recently has the development of a program become in principle a sequence of correctness-preserving refinement steps from its (initial) specification to its (final) code.

Functional programs, however, are just mathematical expressions in disguise, and so for the mathematician the idea of refining them has always been very natural: a sequence of refinement steps corresponds to a chain of equalities in mathematical reasoning. A functional programmer starts with a simple but inefficient program and, through a sequence of refinements, reaches a complex but more efficient one. A mathematician who wanted to paint the area under the famous bell-shaped Normal Curve would first write the specification $\int_{-\infty}^{\infty} e^{-x^2} dx$, and then refine his way to the executable $\sqrt{\pi}$ — executable in the sense that it tells him directly how much paint to buy. Pure LISP, FP, and (pure) ML are examples of functional programming languages.

Concurrent or distributed programs are those in which several components can be active simultaneously. One attraction of such programs is that they are usually faster than their sequential counterparts; and in any case, physical distribution might be part of the client's requirements. But producing these programs systematically — by refinement — is at present still quite difficult in general. And there is no consensus on the method, nor is there a 'typical' programming language for concurrency.

Finally, digital hardware circuits are similar in their effect to programs and are perhaps of even more importance. For a typical circuit may be much more widespread than a typical program; the consequences of its being incorrectly implemented are correspondingly more severe. Worse still, circuit malfunctions are hard to diagnose and even harder to repair. It is fortunate, then, that refinement can work for those as well.

The above, then, are the four applications of refinement covered by this volume, and indeed they are quite different. But the following parable may give the flavour of their common concerns.

The client is a high priest who wants to build a holy temple, and his main concern is that the facade should be of the right size and shape. He calls in a firm of builders, and explains carefully to them the required Sacred Shape. But when the temple is delivered, he discovers that they have not *quite* understood the subtleties of his profession: yes it is of the right height, but unfortunately is just a little too wide... Expensive changes to the stonework are then begun.

Well, perhaps not. A better approach would be to engage at the very beginning an architect for the analysis of requirements. He discovers from the priest that the Sacred Shape is indeed exactly 10 meters high. But its width must give the facade the property illustrated in Figure 1: removing a square leaves the same shape again, but smaller.

Though not a priest himself, the architect has still made the requirements precise in terms that both he and his client can understand. Indeed, that is his job. But it cannot be expected of course that those are terms comprehensible to builders, who as we have seen must be told exactly how wide and how high to build. Thus some refinement is called for.

A correct sequence of refinements will result in building instructions of the form

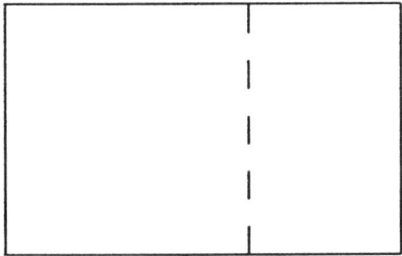

Figure 1: The Sacred Shape

    width   (so-and-so)
    height  (so-and-so)

that imply the property of Figure 1: *if* the temple is so-high and so-wide *then* it will be of the Sacred Shape. The refinement method needed in this case is one that suits the discovery of numbers with certain properties.

Once a refinement method is selected, the requirements must be cast in its terms: always a tricky step, since a mistake here will lose everything. For that reason must those terms be as expressive as possible in order to capture convincingly the informal (but still precise) requirements. The refinement method *elementary algebra* is chosen, and the formal specification is carefully written:

Let $h$ be the height and $w$ be the width, both in meters. Then

$$h = 10$$
$$\text{and} \quad w/h = h/(w - h).$$

See Figure 2.

Now can the refinements begin. The above is *implied by*

$$h = 10 \text{ and } w/10 = 10/(w - 10),$$

and the architect's temple-development team proceeds with the second expression as follows:

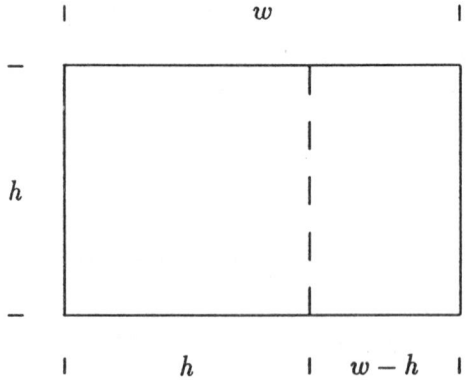

Figure 2: Defining $w$ and $h$

$$\begin{array}{lll} & w/10 = 10/(w-10) & \text{— from above} \\ \text{if} & w^2 - 10w = 100 & \\ \text{if} & w^2 - 10w + 25 = 125 & \text{— complete the square} \\ \text{if} & w - 5 = 5\sqrt{5} & \text{— reduce nondeterminism} \\ \text{if} & w = 5 + 5\sqrt{5}. & \end{array}$$

Note well that the second-last step, whose conclusion is labelled 'reduce nondeterminism', is *not* also 'only if'.

Finally the simple and restricted language of the builders is reached, and the instructions can be given:

width $5 + 5\sqrt{5}$
height $10$.

Those indeed can be directly executed.

The above allegory covers most of the features for the construction of implementations by refinement. The transition from 'the Sacred Shape' — an imprecise notion — to Figure 1 is requirements analysis. The first equation then is the result of translating that into the terms of the chosen refinement method, in this case elementary algebra. The equation makes it clear what property $w$ must have — but it is inefficient in the sense that the actual value is not at all apparent.

The subsequent steps '*this* if *that*' are the refinement steps exactly: and they preserve correctness because in every case if the client is happy with *this* it must be that he is happy with *that* also. The steps are made by developers — good ones, in this case, that know advanced implementation techniques like completing the square.

The language in which the specification and its successive refinements are expressed is a very powerful one — and it is not necessarily executable. One cannot assume that builders can solve quadratic equations. So the refinement process must be directed towards an expression in an executable sublanguage that builders *do* understand: in this case, expressions of the form

$$w = (\text{so-and-so}).$$

Returning now to computers, the executable sublanguage might be might be Pascal, ML, occam, or VLSI. But of which more expressive languages are they a part? For that you must read the papers in this volume. In each case will you encounter the theme set out above: that implementations are produced by taking a specification through a sequence of correctness-preserving refinement steps to executable code.

# Back

As refinement proceeds from high- to low-level descriptions, from expressive to restrictive subsets of a programming language, more and more decisions are made by the developer on the basis of the target architecture. For distributed systems, that is particularly interesting.

Distributed systems differ widely in the way their components communicate, from shared memory through asynchronous message passing to synchronous rendezvous. Even within one communication regime the number and connection scheme of the components can introduce further variation.

It is especially important therefore that such details do not intrude too early in the development of a distributed algorithm from its specification: similarly, one would not expect the large-scale structure of a database to be decided by whether there is a test-and-set instruction on the target computer.

Thus in the development of distributed programs are the benefits of refinement especially evident. The restricted sub-language towards which the refinement steps are directed is one that closely matches the target architecture in that particular case. In another case, subsequently on a different architecture, the refinement steps need be repeated only from the point at which the differences became apparent.

In the following paper, refinements transform the authors' *action systems* gradually into a restricted form that can be executed in occam on transputers.

# Deriving an Occam Implementation of Action Systems

R.J.R. Back *  K. Sere *

**Abstract**

The design of parallel and distributed systems in the event-based action system formalism allows the separation of concerns between the problem to be solved and the target language and machine architecture. The problems involved in executing action systems on distributed architectures is studied in this paper. Our main contribution is the definition of a class of action systems that can be mechanically compiled into occam. The occam-implementation is formally derived using a combination of the refinement calculus and the action systems methodology.

## 1 Introduction

The *action system* approach to model parallel and distributed activity was first proposed by Back and Kurki-Suonio in [7]. The behaviour of parallel and distributed programs is in this formalism described in terms of the actions which processes in the system carry out in co-operation with each other. Several actions can be executed in parallel, as long as the actions do not have any variables in common. The actions are atomic: if an action is chosen for execution, it is executed to completion without any interference from the other actions in the system.

The use of action systems permits the design of the logical behaviour of a system to be separated from the issue of how this system is to be implemented. The latter is seen as a design decision that does affect the way in which the action system is built, but is not reflected in the logical behaviour of the system. A similar separation between logical behaviour and implementation is also made in the UNITY approach by Chandy and Misra [14].

Action systems provide a generalization of the communication mechanisms normally found in programming languages for parallel and distributed programming. Parallel implementation of action systems requires some additional mechanism to enforce atomicity and to schedule the execution of actions. Back and Kurki-Suonio [7] show how a class of action systems can be implemented in CSP with output guards [17], when each action involves at most two processes. Crucial to this implementation is the availability of output guards. Efficient algorithms to implement action systems on broadcasting networks are presented in [1,6,8].

We will in this report identify a class of action systems which can be mechanically compiled into occam [18] and executed in a multiprocessor environment in an efficient way. The programming language occam bears a close resemblance to CSP. The main differences between our implementation and that of [7] stems from to the fact that occam does not support output guards and that we permit any number of processes to participate in an action.

The implementation of multi-process handshaking mechanisms is also considered by Ramesh and Mehndiratta [22]. Chandy and Misra [14] study this problem under the name of committee coordination. This implementation problem is also studied by Bagrodia in [12].

The paper is organized as follows. In section 2 we describe the action system approach to the design of parallel and distributed systems. We then discuss a sequential implementation of action systems in occam. Section 3 introduces the notion of *partitioned action systems* as a way of splitting up action systems for parallel execution. In section 4 we define a class of action systems for which efficient implementations on point-to-point networks are feasible, the so-called *choosers-committers*

---

*Åbo Akademi University, Department of Computer Science, SF-20520 Turku, Finland systems. We also define a very restricted class of *occam-like* action systems, which correspond directly to occam programs.

A general method for refining action systems, *superposition*, is informally described in Section 5. This method was formally described by Back and Kurki-Suonio in [7,9] and variants of it are also treated by Chandy and Misra [14], Bouge and Francez [13] and Katz [20].The appendix shows how to derive this method in the refinement calclulus [2,11], as adapted to the refinement of reactive systems by Back in [4]. In the second part of this section, we show how to transform any chooser-committer system into an equivalent occam-like action system by a sequence of superposition steps. We end in section 6 with some ideas on how to further develop and optimize the implementation.

The syntactic transformation from chooser-committer systems to occam-like action systems and from occam-like action systems to occam programs can be mechanized. Thus the net effect of the derivation is a compiler from action systems into occam, which has been verified correct in the refinement calculus.

We will not discuss the issues of fairness as regards the occam implementation of action systems here in more detail, but hope to return to this specific topic in a later study. Fairness issues in connection with action systems and their implementations are discussed in great detail in [8].

## 2 Action systems

We give an overview of the action system approach to the construction of parallel programs below. More details and examples of this approach can be found in [8,5,10]. We will also briefly describe those aspects of the occam language that are of importance for us here, and describe a simple sequential implementation of action systems in occam.

### 2.1 Action systems formalism

**Action systems** An *action system* $\mathcal{A}$ is an initialized iteration statement

$$\mathcal{A} = |[\text{ var } x; S_0; \text{do } A_1 [\!]\ \ldots\ [\!]\ A_m \text{ od }]| : z$$

on *state variables* $y = x \cup z$. The variables $z$ are the global variables and the variables $x$ are local to $\mathcal{A}$. Each variable is associated with some domain of values. The set of possible assignments of values to the state variables constitutes the *state space* $\Sigma$. The initialization statement $S_0$ assigns initial values to the state variables.

Each action $A_i$ is of the form $g_i \rightarrow S_i$, where the *guard* $g_i$ is a boolean condition and the *body* $S_i$ a sequential statement on the state variables. The guard of action $A$ will be denoted $gA$ and the body $sA$. The state variables referenced in action $A$ will be denoted by $vA$. We say that two actions $A$ and $B$ are *independent*, if $vA \cap vB = \emptyset$, otherwise they are *competing*.

The behaviour of a *sequential action system* is that of Dijkstra's guarded iteration statement [15]

$$S_0;\ \text{do } A_1 [\!]\ A_2 [\!]\ \ldots\ [\!]\ A_m \text{ od}$$

on the state variables. The initialization statement is executed first, after which the do-loop is executed, as long as there are actions $A_i$ that are *enabled* (actions whose guards evaluate to *true*). The action system is said to terminate, if any possible execution of the guarded iteration statement terminates.

A formal definition of the sequential execution model for action systems as a transition system is given in [8], where also fairness of execution is taken into account. We will largely ignore the fairness issues here.

**Example 1: Exchange sort** The action system *ExchangeSort* in Figure 1 will sort $n$ integers in ascending order. It consists of an initialization statement together with $n-1$ sorting actions, $Ex.1, \ldots, Ex.(n-1)$. We indicate replication of declarations, statements and actions by a **for**-clause after the construct.

$ExchangeSort$ :
|[  var $x.i \in integer$  for $i = 1, \ldots, n$;
   $x.i := X_i$  for $i = 1, \ldots, n$;
   do
   [] $x.i > x.(i+1) \rightarrow x.i, x.(i+1) := x.(i+1), x.i$         ($Ex.i$)
   for $i = 1, \ldots, n$;
   od
]|

Figure 1: Exchange sort

Actions $Ex.1$ and $Ex.3$ are e.g. independent, but action $Ex.2$ is competing with both these actions. The action system terminates when the array $x$ is sorted.

**Example 2: Producer–consumer system**  Let us consider a variation of the producer–consumer system. Assume that we have a number of producers which generate items that are stored in a bounded buffer $Q$ with maximum capacity $L$, $L > 0$. The consumers receive items from the same buffer.

An action system describing the behaviour of producers and consumers is shown in Figure 2. The producers are represented by the variables $x.i, i = 1, \ldots, N$ and the consumers by the variables $y.j, j = 1, \ldots, M$. A producer produces only a finite number of items $N_i$. The boolean $prod.i$ denotes whether element $x.i$ contains a produced item or not. It is initialized to $false$. The boolean $cons.j$ denotes correspondingly whether item $y.j$ has been consumed or not. It is initialized to $true$. The action system $ProducerConsumer$ has $2 \times (N + M)$ actions. A $Produce.i$ action produces an item when $prod.i$ equals $false$ and there are items left to produce. Similarly, $Consume.j$ consumes an item when $cons.j$ is $false$. A produced item $x.i$ is stored in the buffer by a $Deposit.i$ action. This can be activated when there is an item in $x.i$ and there is space in the buffer. An item can be removed from the buffer through an $Extract.j$ action. This action is enabled when the buffer is not empty and the previous element $y.j$ has been consumed.

The $Produce.i$ actions are mutually independent and so are the $Consume.j$ actions. Each pair of actions $Produce.i$ and $Consume.j$ are also independent. $Produce.i$ and $Deposit.k$ actions are idenpendent when $i \neq k$, as are $Consume.j$ and $Extract.l$ when $j \neq l$.

The behaviour of the action system is as follows. The initialization enables only the $Produce.i$ actions. The execution of $Produce.i$ enables $Deposit.i$. Each $Deposit.i$ action again enables the corresponding $Produce.i$ action and possibly also $Extract.l$ actions. If an $Extract.l$ action is executed, the corresponding $Consume.l$ action becomes enabled. The action system terminates when $\Sigma_{i=1}^{N} N_i$ items have been produced and consumed.

## 2.2  Sequential action systems in occam

We next discuss our target language occam, concentrating on those features that are important for the implementation. We also give an interpretation of sequential action systems in occam. The language occam is the main language for programming the transputers [19].

**The occam programming language**  An occam program consists of a finite number of processes $P_1, \ldots, P_n$. These processes can be composed in parallel by a $PAR$-construction (Figure 3a) and sequentially by a $SEQ$-construction (Figure 3b). A non-deterministic choice between a number of alternatives is performed in an $ALT$-construction (Figure 3c).

Processes in a parallel composition can only refer to their local variables, so there are no shared variables. Processes communicate by sending and receiving messages. Message–passing in occam is synchronous. Each pair of communicating processes $P_i, P_j$ need to be connected by a directed

$ProducerConsumer$ :
$\|[\ \textbf{var}\ (x.i \in real, n.i \in integer, prod.i \in boolean)\ \textbf{for}\ i = 1, \ldots, n;$
$\quad\quad\quad (y.i \in real, cons.i \in boolean)\ \textbf{for}\ i = 1, \ldots, M;$
$\quad\quad\quad Q \in queue\ of\ real;$
$\quad prod.i, n.i := false, 0\ \textbf{for}\ i = 1, \ldots, N;$
$\quad cons.i := true\ \textbf{for}\ i = 1, \ldots, M;$
$\quad Q := \langle\rangle;$
$\quad \textbf{do}$
$\quad\ \|\ \neg prod.i \wedge n.i < N_i \rightarrow$ $\hfill (Produce.i)$
$\quad\quad\quad produce.item(x.i);\ prod.i := true;\ n.i := n.i + 1$
$\quad\ \textbf{for}\ i = 1, \ldots, N$
$\quad\ \|\ prod.i \wedge |Q| < L \rightarrow$ $\hfill (Deposit.i)$
$\quad\quad\quad Q := Q \cdot (x.i);\ prod.i := false$
$\quad\ \textbf{for}\ i = 1, \ldots, N$
$\quad\ \|\ cons.i \wedge |Q| > 0 \rightarrow$ $\hfill (Extract.i)$
$\quad\quad\quad y.i, Q := head(Q), tail(Q);\ cons.i := false$
$\quad\ \textbf{for}\ i = 1, \ldots, M$
$\quad\ \|\ \neg cons.i \rightarrow$ $\hfill (Consume.i)$
$\quad\quad\quad consume.item(y.i);\ cons.i := true$
$\quad\ \textbf{for}\ i = 1, \ldots, M$
$\quad \textbf{od}$
$]\|$

Figure 2: Producer–consumer system

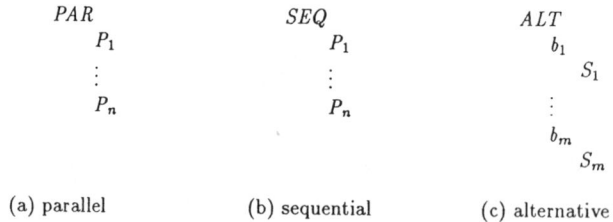

(a) parallel  (b) sequential  (c) alternative

Figure 3: occam constructors

```
SEQ
  S_0
  WHILE gA_1 ∨ ... ∨ gA_n
    ALT
      gA_1
        sA_1
      ⋮
      g_m
        sA_m
```

Figure 4: Sequential occam implementation

communication channel $c$. The input statement $c\ ?\ x$ in process $P_j$ waits for input from process $P_i$ along the channel $c$ and saves the value received in the variable $x$. The matching output statement $c\ !\ e$ in process $P_i$ outputs a value $e$ to process $P_j$ via $c$. The combined effect of these statements is that of the assignment statement $x := e$.

An $ALT$-construction is executed by first evaluating the guards $b_i$. One alternative whose guard $b_i$ evaluates to *true* is then non–deterministically chosen and the corresponding statement $S_i$ is executed. A guard is either a pure boolean guard, an input statement, *SKIP* or a combination of these. The construct aborts if all boolean parts of the guards evaluate to *false*. A *SKIP* guard always evaluates to *true*. An input statement in a guard evaluates to *true* as soon as the corresponding output statement (in some other process) is waiting to be executed. Output statements in guards are not allowed in occam.

**Sequential action systems in occam** An action system can be written directly as a sequential occam program. The action system $\mathcal{A}$ above corresponds to the occam program in Figure 4. As all the action guards are just boolean conditions, the $ALT$-construction evaluates all of them and chooses non–deterministically one enabled action for execution. There is no parallelism in this program.

## 3 Partitioned action systems

Consider an action system

$$\mathcal{A} = |[\ \text{var}\ x; S_0;\ \text{do}\ A_1\ |\ ...\ |\ A_m\ \text{od}\ ]|:z$$

and let $y = x \cup z$. Let $\mathcal{P} = \{p_1, \ldots, p_k\}$ be a partitioning of $y$, i.e.

(i) $p_i \subseteq y$ and $p_i \neq \emptyset$, for $i = 1, \ldots, k$,

(ii) $\bigcup_{i=1}^{k} p_i = y$ and

(iii) $p_i \cap p_j = \emptyset$ when $i \neq j$, for $i, j = 1, \ldots, k$.

The tuple $(\mathcal{A}, \mathcal{P})$ is called a *partitioned action system*.

We identify each partition $p_i$ with a *process*, with the variables in $p_i$ as local variables. We say that action $A$ *involves* process $p$, if $vA \cap p \neq \emptyset$. The local variables of process $p$ referenced in an action $A$ are denoted $vA(p)$.

Let $pA = \{p \in \mathcal{P}\ |\ A\ \text{involves}\ p\}$. Two actions $A$ and $B$ are *independent in partitioning* $\mathcal{P}$ if $pA \cap pB = \emptyset$. An action $A$ is *private* in $\mathcal{P}$, if $|pA| = 1$, otherwise it is *shared* in $\mathcal{P}$. We permit actions that are inependent in some partition to be executed in parallel in that partition. As two independent actions do not have any variables in common, the parallel execution is equivalent to executing the actions one after the other, in either order.

We say an action $A$ is of *degree* $n$ in $\mathcal{P}$ if $|pA| = n$. An action system $\mathcal{A}$ in $\mathcal{P}$ is said to be of *degree* $n$ if all actions are of at most degree $n$.

**Example 1: Exchange sort** Consider the system *ExchangeSort* and the following partitions of the elements in array $x$:
$$\mathcal{P}_1 = \{\{x.1, \ldots, x.n\}\}$$
$$\mathcal{P}_2 = \{\{x.1\}, \ldots, \{x.n\}\}$$
$$\mathcal{P}_3 = \{\{x.1, x.2\}, \ldots, \{x.(n-1), x.n\}\}$$

In partitioning $\mathcal{P}_1$ we only have one process, and all actions are private in this partitioning. In partitioning $\mathcal{P}_2$ we have $n$ processes and each action involves 2 processes. In partitioning $\mathcal{P}_3$ half of the actions are private and the other half involves two partitions.

In partitioning $\mathcal{P}_2$ the actions $Ex.1$ and $Ex.3$ may e.g. execute simultaneously. In partitioning $\mathcal{P}_3$ these actions cannot execute simultaneously, but the actions $Ex.1$ and $Ex.4$ can execute simultaneously. $Ex.1$ is private in the first process while $Ex.4$ involves the second and third process.

**Example 2: Producer–consumer system** Consider the *ProducerConsumer* action system and the partitioning
$$\mathcal{P} = \{\{prod.1, x.1, n.1\}, \ldots, \{prod.N, x.N, n.N\}, \{Q\}, \{cons.1, y.1\}, \ldots, \{cons.M, x.M\}\}.$$

We thus have $N + 1 + M$ partitions: $N$ *Producer*-processes, one *Buffer*-process and $M$ *Consumer*-processes.

The *Produce* and *Consume* actions are private. The *Deposit* and *Extract* actions involve the *Buffer*-process together with one *Producer* respective one *Consumer* process. All *Produce* and *Consume* actions can be executed simultaneously. The *Deposit.k* actions can all executed simultaneously with *Produce.i*, $i \neq k$, and *Consume.j*. The same holds for all *Extract.l* actions with $j \neq l$.

## 4 Parallel action systems in occam

In this section we first show how to detect, in a distributed fashion, the moment when an action becomes enabled. This problem was solved by Back and Kurki–Suonio [7] in the case when the target language is CSP with output guards. We adopt a similar solution here and show how to detect action enabledness in occam. We will identify a very restricted class of actions that correspond directly to occam programs. Another, more general class of action systems will then also be identified, which can be efficiently implemented on point-to-point networks.

### 4.1 Decentralized action systems

Let us consider an arbitrary partitioned action system $(\mathcal{A}, \mathcal{P})$. Let $p \in \mathcal{P}$ be a process. A condition which only refers to local variables of process $p$ is said to be *local* to process $p$. An *action guard* is said to be *separable*, if it is a boolean combination of conditions that are local to the processes involved in the action. An *action* is *separable*, if its guard is separable.

A *partitioned action system* is said to be *separable* or *decentralized* if each of its action is separable.

A separable guard $gA$ can be written in disjunctive normal form $gA^1 \vee \ldots \vee gA^k$, where each disjunct is of the form
$$gA^i = \bigwedge_{p \in pA} gA^i(p).$$

Here $gA^i(p)$ is the conjunction of conditions that are local to process $p$. A condition $gA^i$ of this form is said to be *simple*.

A separable action $A$ can be replaced by actions $A^1, \ldots, A^k$, where each $A^i$ involves the processes $pA$ and is of the form
$$gA^i \to sA$$

$i = 1, \ldots, k$. This collection of actions has the same effect as the original action. Hence, we may in the sequel assume that this action splitting has been done, and that all action guards are simple.

The local conditions $gA^i(p)$ of the simple guard $gA^i$ are called *local guards* of the action, i.e., a simple guard is a conjunction of local guards.

Let $A$ be a decentralized action with $pA = \{p, q\}$, of the form

$$gA(p) \land gA(q) \rightarrow$$
$$y(p), y(q) := f_p(y(p), y(q)), f_q(y(p), y(q)).$$

The effect of executing $A$ can be achieved in CSP by the combined effect of the following two communication statements [7], when output guards are permitted:

$A(p): \quad gA(p) \land q ? y'(q) \rightarrow$
$\qquad\qquad q ! y(p); y(p) := f_p(y(p), y'(q))$

$A(q): \quad gA(q) \land p ! y(q) \rightarrow$
$\qquad\qquad p ? y'(p); y(q) := f_q(y'(p), y(q))$

Here $y'(q)$ and $y'(p)$ denote local copies of $y(q)$ and $y(p)$ in $p$ and $q$, respectively. The use of output guards makes scheduling of actions simple. However, we cannot use this implementation here, as occam does not have output guards.

The implementability criteria in [7,8] is the separability of action guards. This is also our requirement. We are more liberal than [7] in that we permit action systems of any (finite) degree. Instead we need to make some additional restrictions on the class of action systems, to get around the lack of output guards in occam without loosing efficiency.

**Example 1: Exchange sort** The action system $(ExchangeSort, \mathcal{P}_1)$ is decentralized, because we only have one partition: all the variables are local to this partition. In the system $(ExchangeSort, \mathcal{P}_2)$ none of the guards are separable, because we compare values of variables residing in two different partitions. In the third system $(ExchangeSort, \mathcal{P}_3)$ half of the actions are private and thus separable. The other half is not separable.

**Example 2: Producer–consumer system** The action system $(ProducerConsumer, \mathcal{P})$ is a decentralized action system as each guard is separable.

## 4.2 Splitting actions into occam alternatives

Let $A$ be a decentralized action involving the partitions $p$ and $q$ of the form

$$A: \quad gA(p) \land gA(q) \rightarrow x := e,$$

where $x$ only refers to variables in partition $q$ and $e$ only refers to variables in partition $p$. The effect of action $A$ can be achieved by the joint execution of the following two occam alternatives:

$A(p): \quad gA(p)$
$\qquad\qquad c[p, q] ! e$

$A(q): \quad gA(q) \land c[p, q] ? x$
$\qquad\qquad SKIP$

The action is executed when $gA(p) \land gA(q)$ holds and $q$ is willing to receive the value of expression $e$ from $p$ along channel $c[p, q]$.

More generally, consider an arbitrary decentralized action $A$ with $pA = \{p, q_1, \ldots, q_k\}$, $k \geq 0$, of the form

$$A: \quad gA(p) \land \bigwedge_{i=1}^{k} gA(q_i) \rightarrow \qquad\qquad\qquad (1)$$
$$S_1;$$
$$x_1, \ldots, x_k := e_1, \ldots, e_k;$$
$$S_2; T_1; \ldots; T_k$$

where we assume that $x_i$ is in partition $q_i$, $i = 1, \ldots, k$, while the expressions $e_i$ only refer to variables in partition $p$. The statements $S_1, S_2, T_1, \ldots, T_k$ are optional. If present, $S_1$ and $S_2$ refer only to local variables of partition $p$ and each $T_i$ refers only to the local variables of partition $q_i$. This action corresponds to the following occam alternatives:

$$A(p): \quad gA(p)$$
$$S_1$$
$$c[p, q_1] \, ! \, e_1$$
$$\vdots$$
$$c[p, q_k] \, ! \, e_k$$
$$S_2$$

$$A(q_1): \quad gA(q_1) \wedge c[p, q_1] \, ? \, x_1$$
$$T_1$$

$$\vdots$$

$$A(q_k): \quad gA(q_k) \wedge c[p, q_k] \, ? \, x_k$$
$$T_k$$

We will say that the action $A$ is *occam–like* when it can be written in the form (1). We will then refer to $p$ as the *sender* and $q_1, \ldots, q_k$ as the *receivers* of action $A$.

The combined effect of executing these alternatives in occam is the same as executing the action $A$. This follows from the fact that $S_2, T_1, \ldots, T_k$ are independent of each other and hence can be executed in any order.

A private action $A$ ($k = 0$) involving process $p$ becomes just the alternative

$$A(p): \quad gA(p)$$
$$sA.$$

### 4.3 Atomicity and deadlock freedom

The problem with this scheme is that the sender $p$ is locked into action $A$ as soon as $gA(p)$ holds and it enters this alternative. If $A$ does not become enabled after $p$ has decided to choose this alternative, some of the receivers $q_i$ will never be willing to communicate with $p$, i.e., $p$ is stuck in an unexecutable alternative. At the same time there could have been some other action for $p$ to chose which would have become enabled.

On the other hand, if $gA$ is guaranteed to hold whenever $gA(p)$ holds, then this situation cannot occur. Let us make this into a definition.

The enabledness of a separable action $A$ with $pA = \{p, q_1, \ldots, q_k\}$ is said to be *determined* by $p$, if

$$gA(p) \Rightarrow \bigwedge_{i=1}^{k} gA(q_i)$$

holds. The above problem is thus avoided if the sender in an occam-like action always determines the action enabledness.

However, there is still a possibility of deadlock in this scheme. Consider the three actions $A$, $B$ and $C$ with $pA = \{p, q\}, pB = \{q, r\}$ and $pC = \{r, p\}$. Let each action be occam–like with $gA(p) \Rightarrow gA(q), gB(q) \Rightarrow gB(r)$ and $gC(r) \Rightarrow gC(p)$. Assume that all the three actions are simultaneously enabled. If process $p$ chooses alternative $A$, process $q$ chooses alternative $B$ and process $r$ chooses alternative $C$, the three processes are deadlocked in the occam implementation.

Another problem is that the above scheme does not guarantee atomicity when there are more than two processes involved in an action. Consider again action $A$ with $pA = \{p, q_1, \ldots, q_k\}$ and assume that process $p_i$ determines the enabledness of $A$. This does not mean that a process $q_j$ cannot also be involved in some other enabled action, and choose to participate in this instead. In this case, it is possible that some processes $q_j$ receive the values from $p$ and start the action, while some other processes get engaged in some other action instead. In this case, atomicity of action execution is not guaranteed.

To avoid these problems, we need to put additional constraints on the action systems. Consider therefore an arbitrary decentralized action system $(\mathcal{A}, \mathcal{P})$. Let us pick for every action $A$ in $\mathcal{A}$ one of the processes in $pA$ and call it the *chooser* for $A$. The other processes in $pA$ are called *committers*. We denote the unique chooser for $A$ by $cA$. Private actions have a single chooser and no committers.

A decentralized action system $(\mathcal{A}, \mathcal{P})$ is called a *chooser-committer action system* when the following requirement holds for each pair of actions $A$ and $B$, $A \neq B$:

$$gA(p) \Rightarrow \neg gB(p) \quad \forall p \in pA \cap pB, p \neq cA.$$

In these systems each committer is locked to its only locally enabled action. Only a chooser has the freedom to select between different actions for execution.

Let us say that a statement $S$ is *partitioned* among the processes $p_1, \ldots, p_k$, if it is of the form $S = S(p_1); \ldots; S(p_k)$, where $S(p_i)$ only refers to variables in partition $p_i$.

A partitioned action system will be called *occam-like* if the following conditions are satisfied:

(i) The action system is choosers-committers.

(ii) All actions are occam-like.

(iii) The initialization statement is partitioned among the processes in the partitioning.

(iv) Whenever the chooser is also the sender, it determines action enabledness.

(v) The sender and the chooser must be the same in actions involving more than two processes.

This means that different requirements are put on the actions depending on the number of processes involved. There are no restrictions at all for private actions. For two process actions, there are no restrictions if the sender is a committer. If the sender is also the chooser, then it must determine action enabledness. For actions with three or more participants, the chooser and the sender must be the same, and the sender must determine action enabledness.

An occam-like action system can be directly translated into occam as follows. Let the system have $n$ partitions (processes) $p_1, \ldots p_n$. Assume that $S_0 = S_0(p_1); \ldots; S_0(p_n)$ is the initialization statement. Each process $p_i$ executes the following statement $P_i$:

$P_i$ : $SEQ$
$\qquad S_0(p_i)$
$\qquad WHILE\ gA'_1(p_i) \vee \ldots \vee gA'_{m_i}(p_i)$
$\qquad \quad ALT$
$\qquad \qquad A'_1(p_i)$
$\qquad \qquad \vdots$
$\qquad \qquad A'_{m_i}(p_i)$

where $A'_1, \ldots, A'_{m_i}$ are those actions that involve partition $p_i$ and $A'_i(p_j)$ is the alternative in process $p_j$ that corresponds to action $A'_i$, as described above.

The processes $P_i$ are executed in parallel inside an occam $PAR$-construction $P$:

$P$ : $PAR$
$\qquad P_1$
$\qquad \vdots$
$\qquad P_n$

Execution of $P$ in occam corresponds to a parallel execution of the partitioned action system $(\mathcal{A}, \mathcal{P})$.

This gives us an occam implementation of occam-like action systems, where the actions are scheduled and executed in a distributed fashion by the processes involved in it. No extra control messages are needed in this implementation scheme.

**Example 1: Exchange sort** The *ExchangeSort* action system is decentralized only in partitioning $\mathcal{P}_1$. In this partition the system is also a chooser–committer action system as all the actions are private. For the same reason $(ExchangeSort, \mathcal{P}_1)$ is also occam–like.

**Example 2: Producer–consumer system** The *ProducerConsumer* action system is a chooser–committer action system in partition $\mathcal{P}$ when we define the *Buffer*–process to be the chooser of each shared action.

Should we define the *Producer*– and *Consumer*–processes as choosers for these actions, the shared actions do not satisfy the above requirement. We namely have that if e.g. $0 < |Q| < L$ holds, the local guards of the *Buffer*–process for all the *Deposit* and *Extract* actions are enabled.

The *Deposit* actions are occam–like with the producer as the sender. However, the *Extract* actions are not occam–like as the *Buffer*–process is the sender, but it does not determine action enabledness. Recall that this process is the chooser for the *Deposit* actions. Hencs, $(ProducerConsumer, \mathcal{P})$ is not an occam–like action system.

### 4.4 Correctness of implementation

We will show that the implementation of occam–like action systems in occam is correct. The proofs are somewhat informal in nature, as we do not introduce any formal model for occam program executions here. Assume that the occam–like partitioned action system $(\mathcal{A}, \mathcal{P})$ and its occam implementation $P$ are as described in the previous subsection.

The following theorem shows that the occam implementation is safe, in the sense that it does not execute anything else than enabled actions in the corresponding partitioned action system.

THEOREM 1 *The occam implementation $P$ will only execute enabled actions from the occam-like partitioned action system $(\mathcal{A}, \mathcal{P})$.*

*Proof* Consider any process $p$ in the partitioning $\mathcal{P}$. Assume first that $p$ is the sender of an action $A$, that $gA(p)$ holds and that $p$ chooses this alternative, possibly among some other alternatives.

If $A$ is a private action, then process $p$ will execute the action body $sA$ directly, and the action $A$ has been executed atomically.

If $A$ is a two process action and $p$ is a committer for this action, then $gA(p)$ is the only action that $p$ can engage in. Then $p$ will wait until the chooser of the action evaluates its local guard for $A$ to true and also chooses this action. When this happens, then the action $A$ will be executed to completion without interference from other actions, in an atomic fashion. If the chooser never selects this alternative, then either the action would not have become enabled in an execution of the action system or the chooser consistently chooses some other action to participate in. As we do not assume fairness here, the latter possibility is permitted also by the execution of the action system.

The last possibility is that $p$ is the chooser for action $A$, which may involve two or more processes. By assumption, $p$ then determines the enabledness of $A$. Hence, as $gA(p)$ holds, then also the local guards for all the other actions hold. Moreover, all other actions involved in $A$ are committers, so they are not willing to participate in any other action than $A$. Therefore, they will all accept communication from the sender before accepting any other alternative. Consequently, the action will be executed atomically also in this case.

Let us then consider a process $q$ that is a receiver of action $A$ and which evaluates its local guard for participation in $A$ to true. If $A$ is a two process action and $q$ is a chooser for $A$, then it may choose alternative $A(q)$, if the committer for this action (which is also the sender) is ready for communication. In this case the action $A$ is executed atomically to completion by the two processes. Or it may choose some other alternative, for which the same situation then holds. In either case, $q$ will execute some action atomically, if it does execute one of the alternatives that are enabled. If $q$ is again a committer for the action $A$, then this alternative is the only possible for it. Hence, it will wait until the chooser for this action (which is also the sender) determines to select this alternative, in which case the action is executed atomically to completion by these two processes. In case action

$\mathcal{A}$ has more than two participating processes, then the same holds for $q$, as $q$ must then by necessity be a committer.

Thus we see that the occam program $P$ will only execute enabled actions in an atomic fashion in the partitioned action system $(\mathcal{A}, \mathcal{P})$. □

As the above theorem shows, the occam implementation cannot do anything wrong. It is, however, still possible that it might deadlock when there are still actions that could be executed. This possibility is ruled out by the following theorem.

THEOREM 2 *The occam implementation $P$ will not deadlock if there are enabled actions in the partitioned action system $(\mathcal{A}, \mathcal{P})$.*

*Proof* Assume that the occam implementation has deadlocked, but that there is some enabled action $A$.

Assume first that $A$ is an action with two or more processes, where the sender is also the chooser. If the chooser has selected this action, then there cannot be a deadlock, because all the other processes must be willing to participate in this action, as they are committers and their local guards for $A$ must hold because the sender determines the enabledness of the guard. If the chooser has not selected this alternative, then it must be because it has selected some other alternative in which it is also the chooser. But in that case this action can be executed, either because it is a private action which can be executed directly, or it is an action with two or more processes, for which the same argument holds as just stated.

If $A$ is a private action, then the only possibility for deadlock is that the chooser of $A$ is a sender for some other action, as well as a chooser for this action, .i.e., there is another enabled action where the chooser is the sender and which has been selected by the sender, but which is deadlocked. But this is ruled out by the previous argument.

Finally, $A$ could be a two process action where the sender is a committer. Then, as $A$ is assumed to be enabled, the local guard for the chooser for action $A$ must hold, so the only way in which there could be a deadlock is again that the chooser is a sender for some other action, and has selected that. But again, this alternative is not possible. □

Finally, there is a possibility that the occam implementation does not terminate when the action system terminates. This is ruled out by the following theorem.

THEOREM 3 *The occam implementation $P$ terminates in a distributed fashion when the corresponding partitioned action system $(\mathcal{A}, \mathcal{P})$ terminates.*

*Proof* Assume that the action system has terminated, i.e., no action is enabled. As shown above, the occam implementation can only execute enabled actions in an atomic fashion. Hence, no process can select and execute any of its alternatives, because then some enabled action would have to be executed jointly by the processes. Hence, when no action guard is enabled, no alternative can be chosen in any the occam processes, i.e. the program $P$ has terminated in a distributed fashion. □

# 5 An occam implementation of action systems

In this section we show how to implement action systems in occam. We first describe, somewhat informally, a general program transformation rule, superposition. This rule is then used in the second subsection repeatedly to construct an occam implementation of choosers–committers action systems, by a sequence of correctness preserving refinement steps.

## 5.1 Superposition

Program refinements very often proceed by adding new variables to the state, together with code that manipulates these new variables. The purpose is often to avoid recomputing values, or, in parallel programs, to distribute the state over many processors. When the addition of new variables and associated computation code is done in a way that preserves the old computation of the program from being disturbed, then we call it *superpositioning*. This method for program refinement has come up in a number of different contexts, e.g. in the works of Dijkstra and al. [16], Back and Kurki–Suonio [7], Chandy and Misra [14], Bouge and Francez [13], Katz [20] and so on.

The superposition method can be derived as a special case of the method for data refinement in the refinement calculus, described by Back in [3] and extended by Back and vonWright in [11], in combination with the method for refinement of reactive programs described by Back in [4]. The formal derivation of this method is shown in the appendix, and assumes that the reader is familiar with [11,4]. For the derivation in the next section, an informal description of this method as given below, is sufficient.

Let $|[\textbf{ var } v; S_0; \textbf{do } A_1 \mathbin{\|} \ldots \mathbin{\|} A_m \textbf{ od }]| : x$ be an action system. Superposition means modifying this by

(a) adding some new local variables $x$,

(b) adding an initialization of these new variables,

(c) changing the old actions $A_i$ to also update the new variables and

(d) adding some totally new actions, to update the values of the new variables.

These changes should be done in such a way that the effect of the new action system on the old variables is essentially the same as the effect of the old action system on these variables.

**Superposition refinement rule** We have the following general result about superposition (proof in appendix A).

The refinement

$$|[\, S_0; \textbf{do } A_1 \mathbin{\|} \ldots \mathbin{\|} A_m \textbf{ od }\,]| : v$$
$$\leq\ |[\,\textbf{var } x; S_0'; \textbf{do } A_1' \mathbin{\|} \ldots \mathbin{\|} A_m' \mathbin{\|} B_1 \mathbin{\|} \ldots \mathbin{\|} B_n \textbf{ od }\,]| : v$$

holds if the following conditions are satisfied for some invariant $I(v, x)$:

(1) *Initialization:*

   (a) The new initialization $S_0'$ has the same effect on the old variables $v$ as $S_0$ and

   (b) it will establish $I(v, x)$.

(2) *Old actions:*

   (a) The body of each new action $A_i'$ has the same effect on the old variables $v$ as the corresponding old action $A_i$ when $I(v, x)$ holds,

   (b) each new action $A_i'$ will preserve $I(v, x)$ and

   (c) the guard of each new action $A_i'$ implies the guard of the corresponding old action $A_i$, when $I(v, x)$ holds.

(3) *Auxiliary actions:*

   (a) None of the auxiliary actions $B_j$ has any effect on the old variables $v$ when $I(v, x)$ holds,

   (b) each auxiliary action $B_j$ will preserve $I(v, x)$.

(4) *Termination of auxiliary actions:* Executing only auxiliary actions in an initial state where $I(v, x)$ holds will necessarily terminate.

(5) *Exit condition:* The exit condition of the new action system implies the exit condition of the old action system, whenever $I(v, x)$ holds.

The result extends to reactive refinement as well, i.e.,

$$|[\text{ var } v; S_0; \text{do } A_1 \parallel \ldots \parallel A_m \text{ od }]| : z$$
$$\preceq |[\text{ var } v, x; S_0'; \text{do } A_1' \parallel \ldots \parallel A_m' \parallel B_1 \parallel \ldots \parallel B_n \text{ od }]| : z.$$

will hold under the same conditions as above, with the restriction that the invariant $I$ and the auxiliary actions $B_j$ may only refer to the variables $v$ and $x$. This means that the latter action system may replace the former also in a reactive context, in which the variables $z$ may also be updated by other actions in parallel action systems.

A number of very useful program refinement rules arise as special cases of the general superposition rule. We have a simpler rule for the case when we do not add any new auxiliary actions, but only modify existing actions. In that case, we only need to prove properties (1), (2) and (5). An even simpler rule does not even introduce any new variables, but only establishes an invariant and refines action bodies and guards using this invariant. Then we need not prove properties (1a) and (2a). A final simplification does not even establish the invariant, giving us the usual refinement rule for loops.

## 5.2 Derivation of occam–like action systems

We will now show that every chooser–committer action system can be transformed into an equivalent occam–like action system. The derivation makes use of the superposition method throughout, as described above.

Let $\mathcal{A}_0$ be an arbitrary chooser–committer action system with the partitioning $\mathcal{P} = \{x.r \cup z.r \mid r = 1, \ldots, n\}$:

$$\mathcal{A}_0 : |[ \text{ var } x.r \in T_r \text{ for } r = 1, \ldots, n;$$
$$S_0;$$
$$\text{do}$$
$$\parallel A_i$$
$$\text{for } i = 1, \ldots, m \qquad *$$
$$\text{od}$$
$$]|: z_r \text{ for } r = 1, \ldots, n$$

For each action $A_i$, let the chooser be $cA_i = x.p_i \cup z.p_i$ and the committers be $x.q \cup z.q, q \in Q_i$, i.e., $p_i$ identifies the chooser and $Q_i$ identifies the set of committers of action $A_i$.

**Notation** We will indicate in the text of each program version on the *right hand side* those statements that will be changed in the next version. We write $*$ for a modification of the statement. Similarly, we indicate on the *left hand side* those statements that have changed from the previous version. We write $*$ for a modified statement and $+$ for a new statement that has been added.

We will frequently abbreviate a list of assignments $a.1, \ldots, a.k := h.1, \ldots, h.k$ to simply $a := h$, it being understood that the assignments are made pointwise. Similarly, we write $a$ for $a.1 = true \land \ldots \land a.k = true$ and $\neg a$ for $a.1 = false \land \ldots \land a.k = false$ when $a$ is a list of booleans.

**Introducing willingness announcements** We apply the superposition method by introducing for each action $A_i$ local boolean variables $wA_i.q, q \in Q_i$. The new booleans are assigned to the choosers partition and are initialized to *false*. The purpose of the booleans is to keep track of which committers are willing to execute an action.

We also introduce for each action $A_i$ corresponding boolean variables $nA_i.q, q \in Q_i$, but place each $nA_i.q$ in the partitioning of process $q$. These booleans are also initialized to *false*. They are used to indicate whether the committer $q$ already has indicated its willingness to participate in action $A_i$.

We transform $\mathcal{A}_0$ to the equivalent system $\mathcal{A}_1$ by adding new actions and assignments to the newly introduced variables, while at the same time also strengthening the guards of the old actions and adding assingments to the new variables in these. We will preserve the invariance of

$$wA_i.q = nA_i.q \land (wA_i.q \Rightarrow gA_i(q)),$$

for $i = 1, \ldots, m, q \in Q_i$.

The new action system is as follows.

$$\begin{aligned}
\mathcal{A}_1 : \ &|[ \ \textbf{var } x.r \in T_r \ \textbf{for } r = 1, \ldots, n; \\
+ \ & \quad\quad wA_i.q, nA_i.q \in boolean \ \textbf{for } i = 1, \ldots, m, q \in Q_i; \\
& \quad S_0; \\
+ \ & \quad wA_i, nA_i := false, false \ \textbf{for } i = 1, \ldots, m; \\
& \quad \textbf{do} \\
+ \ & \quad [\!] \ gA_i(q) \land \neg wA_i.q \land \neg nA_i.q \land \neg com_i \rightarrow \qquad\qquad * \\
& \qquad\quad wA_i.q, nA_i.q := true, true \\
& \quad \textbf{for } i = 1, \ldots, m, q \in Q_i \\
* \ & \quad [\!] \ gA_i(p_i) \land wA_i \land nA_i \rightarrow \\
& \qquad\quad sA_i; wA_i, nA_i := false, false \qquad\qquad * \\
& \quad \textbf{for } i = 1, \ldots, m \\
& \quad \textbf{od} \\
& \ ]\!|: z.r \ \textbf{for } r = 1, \ldots, n.
\end{aligned}$$

The basic idea is that the auxiliary actions are used to record that a committer is willing to participate in a certain action. The original action will only be enabled when all committers have announced their willingness to participate in it and when the chooser for the action is also willing to particiapate in it.

The condition $\neg com_i$ below has been added to the auxiliary actions to preserve the chooser–committer property. It essentially prevents a process from accepting willingness messages from a committer for an action if the process can itself commit to another action. We define

$$com_i = \bigvee_{j : p_i \in Q_j} gA_j(p_i),$$

i.e., $com_i$ is true if $p_i$ is a committer for some other action $A_j$ and its local guard for this action is enabled.

This refinement satisfies the superposition conditions:

(1) The initialization of the new variables does not affect the initialization of the old variables and they establish the required invariant.

(2) We only add assignments to new variables in the bodies of the old actions $A_i$, these assignments preserve the invariant, and the guards of the new actions imply the guards of the old actions when the invariant holds.

(3) The auxiliary actions do not affect the values of the old variables, and they also preserve the invariant.

(4) Executing only auxiliary actions will necessarily terminate.

(5) The exit condition of the new action system implies the exit condition of the old when the invariant holds.

**Distributing action bodies** Let us now turn our attention to the action bodies. We have to make the actions occam–like, so the bodies $sA_i$ must be distributed. This is again done by superposition. For each action $A_i$ we add new communication variables $yA_i.p_i$ and $yA_i.q, q \in Q_i$. These variables will keep copies of the state variables, so that $yA_i.p_i$ is a copy of $x.p_i$ and $yA_i.q$ is a copy of $x.q$, $q \in Q_i$. The new variables are all placed in the choosers $p_i$ partition. This gives us the refinement $\mathcal{A}_1 \preceq \mathcal{A}_2$, where

$\mathcal{A}_2 :$ |[ **var** $x.r \in T_r$ **for** $r = 1, \ldots, n;$
    $wA_i.q, nA_i.q \in boolean$ **for** $i = 1, \ldots, m, q \in Q_i;$
  $+$ $yA_i.r \in T_r$ **for** $i = 1, \ldots, m, r \in pA_i;$
  $S_0;$
  $wA_i, nA_i := false, false$ **for** $i = 1, \ldots, m;$
  **do**
* [ $gA_i(q) \wedge \neg wA_i.q \wedge \neg nA_i.q \wedge \neg com_i \rightarrow$
      $wA_i.q, nA_i.q := true, true; yA_i.q := x.q$
    **for** $i = 1, \ldots, m, q \in Q_i$
  [ $gA_i(p_i) \wedge wA_i \wedge nA_i \rightarrow$
  $+$ $yA_i.p_i := x.p_i;$
      $sA_i; wA_i, nA_i := false, false$                  *
    **for** $i = 1, \ldots, m$
  **od**
  ]|: $z.r$ **for** $r = 1, \ldots, n$

The justification for this step is that new local variables are added with assignments only in action bodies.

We next add variables $uA_i.q.r, i = 1, \ldots, m, q \in Q_i, r \in pA_i$. The variable $uA_i.q.r$ will keep a copy of the variable $x.r, r \in pA_i$, in partition $q$. The new action system $\mathcal{A}_3$ is as follows:

$\mathcal{A}_3 :$ |[ **var** $x.r \in T_r$ **for** $r = 1, \ldots, n;$
    $wA_i.q, nA_i.q \in boolean$ **for** $i = 1, \ldots, m, q \in Q_i;$
    $yA_i.r \in T_r$ **for** $i = 1, \ldots, m, r \in pA_i;$
  $+$ $uA_i.q.r \in T_r$ **for** $i = 1, \ldots, m, q \in Q_i, r \in pA_i;$
  $S_0;$                                                    *
  $wA_i, nA_i := false, false$ **for** $i = 1, \ldots, m;$    *
  **do**
  [ $gA_i(q) \wedge \neg wA_i.q \wedge \neg nA_i.q \wedge \neg com_i \rightarrow$  *
      $wA_i.q, nA_i.q := true, true; yA_i.q := x.q$
    **for** $i = 1, \ldots, m, q \in Q_i$
  [ $gA_i(p_i) \wedge wA_i \wedge nA_i \rightarrow$
      $yA_i.p_i := x.p_i;$
  $+$ $uA_i.q := yA_i$ **for** $q \in Q_i;$
  * $sA_i(r)$ **for** $r \in pA_i;$                         *
      $wA_i, nA_i := false, false$                          *
    **for** $i = 1, \ldots, m$
  **od**
  ]|: $z.r$ **for** $r = 1, \ldots, n$

The justification for this step is that

$$wA_i.q \Rightarrow yA_i.q = x.q,$$

is an invariant of the loop, for $i = 1, \ldots, m, q \in Q_i$. Hence the action bodies of the main actions can be refined as follows:

$\{yA_i.q = x.q$ **for** $q \in Q_i\}$
$yA_i.p_i := x.p_i;$
$uA_i.q := yA_i$ **for** $q \in Q_i;$
$sA_i; wA_i, nA_i := false, false$
$\leq$
$yA_i.p_i := x.p_i;$
$uA_i.q := yA_i$ **for** $q \in Q_i;$
$sA_i(r)$ **for** $r \in pA_i;$
$wA_i, nA_i := false, false$

where

$$sA_i(p_i) = sA_i[yA_i/vA_i]; x.p_i := yA_i.p_i$$
$$sA_i(q) = sA_i[uA_i.q/vA_i]; x.q := uA_i.q.q \text{ for } q \in Q_i$$

**Final version** We finally split up the initialization among the processes. The statement $wA_i := false$ is also moved earlier, which is permitted by statement commutativity. This gives us the final program version $\mathcal{A}_4$.

$\mathcal{A}_4 :\ \|[\ \text{var } x.r \in T_r \text{ for } r = 1, \ldots, n;$
$\quad\quad wA_i.q, nA_i.q \in boolean \text{ for } i = 1, \ldots, m, q \in Q_i;$
$\quad\quad yA_i.r \in T_r \text{ for } i = 1, \ldots, m, r \in pA_i;$
$\quad\quad uA_i.q.r \in T_r \text{ for } i = 1, \ldots, m, q \in Q_i, r \in pA_i;$
$\quad * \ [S_0(r);$
$\quad * \ (wA_i := false \text{ for } i : cA_i = r);$
$\quad * \ (nA_i.r := false \text{ for } i : r \in pA_i)]$
$\quad\quad \text{for } r = 1, \ldots, n;$
$\quad \text{do}$
$\quad \|\ gA_i(q) \wedge \neg wA_i.q \wedge \neg nA_i.q \wedge \neg com_i \rightarrow \quad\quad\quad\quad (Commit.i.q)$
$\quad\quad wA_i.q := true; yA_i.q := x.q; nA_i.q := true;$
$\quad\quad \text{for } i = 1, \ldots, m, q \in Q_i$
$\quad \|\ gA_i(p_i) \wedge wA_i \wedge nA_i \rightarrow \quad\quad\quad\quad\quad\quad\quad\quad\quad (Execute.i)$
$\quad\quad yA_i.p_i := x.p_i;$
$\quad\quad uA_i.q := yA_i \text{ for } q \in Q_i;$
$\quad * \ sA_i(p_i); wA_i := false;$
$\quad * \ (sA_i(q); nA_i(q) := false) \text{ for } q \in Q_i;$
$\quad\quad \text{for } i = 1, \ldots, m$
$\quad \text{od}$
$\ ]|: z.r \text{ for } r = 1, \ldots, n$

This action system is now occam–like. Let us consider each of the requirements for this in turn.

(i) The first requirement is that the system is choosers–committers. Consider first action $Commit.i.q$. Here $q$ is committer, for the following reasons. Assume that $q$ is locally enabled in this action, i.e., that $gA_i(q) \wedge \neg nA_i.q$. Then

  (a) $q$ cannot be locally enabled in any other $Commit.j.q$ action, $i \neq j$, because then $gA_i(q) \wedge gA_j(q)$ would hold, contradicting the chooser–committer assumption,

  (b) $q$ cannot be locally enabled as a chooser in any action $Commit.j.p$, $i \neq j$, because $\neg gA_i(q)$ follows from $\neg com_j$,

  (c) $q$ cannot be locally enabled in action $Execute.i$, because $\neg nA_i.q$ holds, and

  (d) $q$ cannot be locally enabled in action $Execute.j$, $i \neq j$, because then $nA_j.q$ must hold, which in turn implies that $gA_j(q)$ would hold, contradicting the chooser–committer property.

  Consider then the action $Execute.i$. Here $p_i$ is the chooser and each process in $Q_i$ is a committer. Assume that $q \in Q_i$ is locally enabled in this action, i,e,, $nA_i.q$ holds. Then $q$ cannot be locally enabled in $Commit.i.q$. It is also not possible that $q$ could be locally enabled in some other action $Commit.j.q$, as, by the invarian, that would again violate the chooser-commiter assumption. Also $q$ cannot be locally enabled in any other $Execute.j$ action, $i \neq j$, for the same reason. Hence we conclude that the choosers–committers restriction is satisfied.

(ii) All actions are occam–like, as required. In action $Commit.i.q$ process $q$ is the sender. In action $Execute.i$ process $p_i$ is the sender.

(iii) The initialization statement is partitioned among the processes as required.

(iv) The chooser is the same as the sender in actions $Execute.i$. In this case the process $p_i$ does indeed determine the enabledness of the action, because $wA_i \Rightarrow nA_i.q$ for every $q \in Q_i$.

(v) The action $Execute.i$ may involve more than two processes, but the sender and the chooser is the same in this action.

Hence, we conclude that the above implementation is an occam–like action system, so it can be translated directly into occam. The derivation establishes the correctness of this implementation.

# 6 Conclusions

We have shown how a class of action systems can be mechanically converted into occam. The restrictions that we have imposed on action systems are basically the same that are needed for efficient implementation of occam itself, i.e. the prohibition of output guards, here expressed more generally as the requirement that the action system must be a choosers–committers system. The implementation allows multi–party actions, and the handshake is itself symmetric, with information being shared between the processes participating in an action in a symmetric way. The chooser–committer restriction permits action systems to be executed on point–to–point networks with an efficiency comparable to occam implementations, as the implementation does not introduce any heavy distributed protocol to schedule actions.

The implementation as described above does not handle termination in the way that would be most convenient for executing multiprocessor algorithms. The occam implementation will deadlock rather than terminate when the action system itself would terminate. Proper termination can be enforced by superimposing a termination detection protocol upon the occam–implementation, in a rather straightforward fashion. A bibliography of methods to detect distributed termination can be found in [21].

We have also not discussed channel assignments on real hardware. On the level of occam–processes we can have any channels between two processes we need. When mapping a process net onto a transputer network, there might be restrictions on the communication between two processes on different transputers. In such a case we must either use routing software or transform the action system further so that only channels which can be mapped directly onto existing hardware channels are used. This latter approach is studied in more detail by Sere in [23].

The code of the implementation can be optimized quite heavily. In the implementation described here, each process executes a copy of the action body. In most cases, this is clearly unnecessary, as the process only needs to find out the new values for its own local variables. In case the local variables are not changed, then it need not execute the action body at all. We have not considered this aspect of the implementation in more detail, as it can be handled by rather standard code analysis methods.

A more difficult optimization problem might be the efficient evaluation of local guards in the processes. The transformation produces a flat occam loop for each process, where all the guards have to be evaluated anew at each iteration, and hence quite a lot of redundant evaluations may be done. The resulting code could be optimized by changing the flat structure into an equivalent decision tree structure where redundant condition testing has been eliminated. Alternatively, one could try to use methods from production system implementations where the same problem occurs, for efficiently determining the set of enabled conditions.

An optimization that would need to be done is to introduce a separate scheduler process for each chooser, which takes care of the communication with the committers. As the implementation now stands, the committers that want to announce their willingness to the chooser will have to wait until the chooser is ready to receive their messages. When the chooser is itself a committer to some other action, it cannot receive these messages. Hence, it is possible that chains of committers build up where nothing happens, while a separate scheduling process could record the willingness messages from the committers while the chooser is waiting.

The implementation permits us to use action systems as a high-level language for programming MIMD-type multiprocessor systems. We are currently developing a compiler from choosers–committers systems to occam along the lines of the implementation. To be really practical as a programming language, one would need to have a more syntactic characterization of choosers–committers systems. By making stronger restrictions, such syntactic characterizations can also be made.

A class of *locally scheduled* action system was defined in [10]. The chooser–committers restriction is stronger that the requirement of being locally scheduled. However, locally scheduled action systems can be implemented by the same method as described here, with a somewhat more elaborate distributed (but still local) protocol.

## Acknowledgements

The work reported here was supported by the FINSOFT III program sponsored by the Technology Development Centre of Finland. We would like to thank Mats Aspnäs, Viking Högnäs, Joost Kok, Reino Kurki-Suonio, David Scillicorn and Marina Waldén, for helpful discussions on the topics treated here.

## A  Refinement of action systems

We give a short summary of the refinement calculus in this appendix. A more complete account of the method for stepwise refinement of parallel algorithms is given in [10], and the extension of this method to stepwise refinement of reactive system is described in [4].

### A.1  Refinement calculus

The *refinement calculus* is a formalization of the stepwise refinement approach for systematic construction of sequential programs. A statement $S$ is said to be *(correctly) refined* by statement $S'$, denoted $S \leq S'$, if

$$\forall P, Q : P[S]Q \Rightarrow P[S']Q.$$

Here $P[S]Q$ stands for the total correctness of $S$ w.r.t. precondition $P$ and postcondition $Q$. This can also be characterized in terms of the weakest preconditions [15] as

$$\forall Q : \mathrm{wp}(S, Q) \Rightarrow \mathrm{wp}(S', Q).$$

The refinement relation is reflexive and transitive. Hence, if we can prove that

$$S_0 \leq S_1 \leq \ldots \leq S_{n-1} \leq S_n,$$

then

$$S_0 \leq S_n.$$

This models the successive refinement steps in a program development: $S_0$ is the initial high level specification statement and $S_n$ is the final executable and efficient program that we have derived through the intermediate program versions $S_1, \ldots, S_{n-1}$. Each refinement step preserves the correctness of the previous step, so the final program must preserve the correctness of the original specification statement.

The refinement relation is monotonic w.r.t. the usual statement constructors. For any sequential statement $S(T)$ that contains $T$ as a substatement, we thus have that

$$T \leq T' \Rightarrow S(T) \leq S(T').$$

It is possible that the replacement of $T$ by $T'$ would preserve correctness in the specific context $S(.)$ where it is used, i.e. $S(T) \leq S(T')$, although the replacement would not be correct in every possible context. This kind of *context dependent replacement* can be handled by the following technique. Let $\{Q\}$ be an *assert statement*, i.e., a statement that acts as a *skip*-statement if $Q$ holds and as an *abort*-statement otherwise. Assume that we can prove

(1) $S(T) \leq S(\{Q\}; T)$ and

(2) $\{Q\}; T \leq T'$.

Then by monotonicity and transitivity we have that $S(T) \leq S(T')$.

Refinement between actions is defined in the same way as refinement between statements, i.e. an action $A$ is *refined by* an action $A'$, $A \leq A'$, if

$$\text{wp}(A, R) \Rightarrow \text{wp}(A', R)$$

for any postcondition $R$. The notion of total correctness can be directly extended to actions: we define $P\langle A \rangle Q$ to hold if $P \Rightarrow \text{wp}(A, Q)$ ($A$ establishes $Q$ when $P$). This is equivalent to $P \wedge gA \Rightarrow \text{wp}(sA, Q)$.

Observe that even if refinement of statements is monotonic, as stated above, action systems are *not* necessarily monotonic with respect to refinement of actions, i.e. $A_i \leq A'_i, i = 1, \ldots, m$, need not imply

$$\text{do } A_1 \ [\!] \ \ldots \ [\!] \ A_m \text{ od} \leq \text{do } A'_1 \ [\!] \ \ldots \ [\!] \ A'_m \text{ od}.$$

However, this implication does hold if we, in addition, require that

$$\bigvee_{i=1}^{m} gA_i \Rightarrow \bigvee_{i=1}^{m} gA'_i$$

(this actually means that the exit conditions for the two loops must be equivalent).

## A.2 Superposition

We write $|[\text{ var } x : I : S ]|$ for the statement $|[\text{ var } x; I \to S; \{I\} ]|$. Hence, this is a block with local variables where it is assumed that the local variables $x$ are assigned initial values that satisfy the invariant $I$, and it is checked that the invariant is still true at the point of exit. A miracle will occur if no value of $x$ can satisfy $I$ and abortion occurs if $I$ does not hold before exit from the block.

The superposition principle is expressed by the following theorem.

THEOREM 4 *Let $\mathcal{A}$ denote the action system* $|[ S_0; \text{do } A_1 \ [\!] \ \ldots \ [\!] \ A_m \text{ od} ]| : v$ *and $\mathcal{A}'$ the action system* $|[\text{ var } x; S'_0; \text{do } A'_1 \ [\!] \ \ldots \ [\!] \ A'_m \ [\!] \ B_1 \ [\!] \ \ldots \ [\!] \ B_n \text{ od} ]| : v$. *Let $gA$ be the disjunction of the guards of the $A_i$ actions, $gA'$ the disjunctions of the $A'_i$ actions and $gB$ the disjunction of the $B_j$ actions. Then $\mathcal{A} \leq \mathcal{A}'$ if the following conditions hold, for some invariant $I(v, x)$:*

(1) $S_0 \leq |[\text{ var } x; S'_0; \{I\} ]|$.

(2) $A_i \leq |[\text{ var } x : I : A'_i ]|$, for $i = 1, \ldots, m$.

(3) $\text{skip} \leq |[\text{ var } x : I : B_j ]|$, for $j = 1, \ldots, n$.

(4) $I[\text{do } B_1 \ [\!] \ \ldots \ [\!] \ B_n \text{ od}] true$.

(5) $I \wedge gA \Rightarrow (gA' \vee gB)$.

*Proof* The proof is based on the proof rule for data refinement in [4, Theorem 1 and Lemma 10][1]. We have by Lemma 10 that $\mathcal{A} \equiv \mathcal{A}^+$, where

$$\mathcal{A}^+ = |[\text{ var } h; S_0; h := h.true; \text{do } A_1; h := h.true \ [\!] \ \ldots \ [\!] \ A_m; h := h.true \ [\!] \ H \text{ od} ]|,$$

where $H = h > 0 \to h := h'.(h' < h)$ is the stuttering action. Hence, it is sufficient to show that $\mathcal{A}^+ \leq \mathcal{A}$. By Theorem 1, we have to show that there exists an encoding statement $\alpha : v, h \to v, x$ such that the following conditions are satisfied:

---

[1] The encoding function $E : x \to x'$ is unnecessarily restricted there, we can as well permit $E : x, z \to x, z$ for total correctness refinement

(i) $\langle \wedge + h \rangle; S_0; h := h.true \leq \langle \wedge + x \rangle; S_0, \alpha^{-1}$,

(ii) $A^+ \leq_\alpha A' \wedge B$, and

(iii) $gA \vee h > 0 \Rightarrow \alpha(gA' \vee gB)$.

Here $A^+ = A_1; h := h.true \wedge \ldots \wedge A_m; h := h.true \wedge H$, $A' = A'_1 \wedge \ldots \wedge A'_m$ and $B = B_1 \wedge \ldots \wedge B_n$.

Let us define for any state $v, x$ the function $b(x, v)$ to be the maximum number of iterations that is needed for the iteration statement do $B_1 \mathbin{[\![} \ldots \mathbin{[\![} B_n$ od to terminate. We assume bounded nondeterminism, so the value of $b(v, x)$ is defined whenever the condition $I(v, x)$ holds, by property (4).

We now define an abstraction relation by

$$R(x, h, v) \stackrel{\text{def}}{=} I(v, x) \wedge h = b(v, x).$$

We need to show that the conditions (i) – (iii) hold with this abstraction relation. The encoding function used is thus $\alpha = \langle \wedge + x - h.R(x, h, v) \rangle$ and the decoding function is $\alpha^{-1} = \langle \vee + h - x.R(x, h, v) \rangle$.

(i) By property (1), we have that $S'_0$ will establish $I$, besides being a refinement of $S_0$ on the variables $v$. Hence, there exists a value $h = b(v, x)$ in the state following the initialization $S'_0$, as required. This establishes the first condition.

(ii) By assumption (2), $A_i; h := h.true \leq_\alpha A'_i$, for each $i = 1, \ldots, m$, because the action $A'_i$ will establish the invariant $I$, and hence that there exists a value $h = b(v, x)$.

Also, we have $H \leq_\alpha B_j$ for each $j = 1, \ldots, n$. This is seen as follows. Assume that in an intial state where the abstraction relation holds, the guard of $B_j$ is enabled. This means that the $B$-loop can be executed at least once, so $h = b(v, x) > 0$. Also, executing the $B_j$ action will necessary decrease the maximum number of iterations of the $B$-loop, which is also well-defined, because $B_j$ preserves the invariant $I$. Also, by assumption, the action $B_j$ does not affect the variables $v$. Thus condition (ii) is seen to hold.

(iii) We finally need to prove that $R(x, h, v) \wedge (gA \vee h > 0) \Rightarrow (gA' \vee gB)$. If $h > 0$, then obviously some action $B_j$ must be enabled, i.e. $gB$ must hold. If again $h = 0$ and $gA$ holds, then no $B_j$ action can be enabled. But by (5), this then implies that $gA'$ must hold. This establishes condition (iii). □

The same proof will establish the following corollary for reactive programs:

COROLLARY 1 *Let the notation and assumptions be as above. Assume that the auxiliary actions $B_j, j = 1, \ldots, n$ and the invariant $I$ only refer to the variables $v, x$. Then we have that*

$$\begin{aligned} & |[\text{ var } v; S_0; \text{do } A_1 \mathbin{[\![} \ldots \mathbin{[\![} A_m \text{ od }]| : z \\ \leq \; & |[\text{ var } v, x; S'_0; \text{do } A'_1 \mathbin{[\![} \ldots \mathbin{[\![} A'_m \mathbin{[\![} B_1 \mathbin{[\![} \ldots \mathbin{[\![} B_n \text{ od }]| : z. \end{aligned}$$

*under the same conditions (1) – (5) as above.*

Proof By the restrictions on the auxiliary variables and the invariant, the encoding function will only refer to the local variables of the block and thus be of the form required for reactive refinement in [4, Theorem 3]. □

# References

[1] M. Aspnäs, R. J. R. Back, and R. Kurki-Suonio. Efficient implementations of multi-process handshaking of broadcasting networks. Reports on computer science and mathematics 75, Åbo Akademi, 1989.

[2] R. J. R. Back. *Correctness Preserving Program Refinements: Proof Theory and Applications*, volume 131 of *Mathematical Center Tracts*. Mathematical Centre, Amsterdam, 1980.

[3] R. J. R. Back. Changing data representation in the refinement calculus. In *21st Hawaii International Conference on System Sciences*, January 1989. Also available as Åbo Akademi reports on computer science and mathematics no. 68, 1988.

[4] R. J. R. Back. Refinement calculus II: Parallel and reactive programs. To appear in the proceedings of the REX Workshop on Stepwise Refinement of Distributed Systems: Models, Formalisms, Correctness, 1989.

[5] R. J. R. Back. Refining atomicity in parallel algorithms. In *PARLE Conference on Parallel Architectures and Languages Europe*, volume 366 of *Lecture Notes in Computer Science*, Eindhoven, the Netherlands, June 1989. Springer Verlag. Also available as Åbo Akademi reports on computer science and mathematics no. 57, 1988.

[6] R. J. R. Back, E. Hartikainen, and R. Kurki-Suonio. Multi-process handshaking on broadcasting networks. Reports on computer science and mathematics 42, Åbo Akademi, 1985.

[7] R. J. R. Back and R. Kurki-Suonio. Decentralization of process nets with centralized control. In *2nd ACM SIGACT-SIGOPS Symp. on Principles of Distributed Computing*, pages 131–142. ACM, 1983.

[8] R. J. R. Back and R. Kurki-Suonio. Distributed co-operation with action systems. *ACM Transactions on Programming Languages and Systems*, 10:513–554, October 1988. Previous version in Åbo Akademi reports on computer science and mathematics no. 34, 1984.

[9] R. J. R. Back and R. Kurki-Suonio. Decentralization of process nets with centralized control. *Distributed Computing*, 3(2):73–87, 1989. Appeared previously in *2nd ACM SIGACT-SIGOPS Symp. on Principles of Distributed Computing 1983*.

[10] R. J. R. Back and K. Sere. Refinement of action systems. In *Mathematics of Program Construction*, volume 375 of *Lecture Notes in Computer Science*, Groningen, The Netherlands, June 1989. Springer-Verlag.

[11] R. J. R. Back and J. von Wright. Refinement calculus I: Sequential nondeterministic programs. To appear in the proceedings of the REX Workshop on Stepwise Refinement of Distributed Systems: Models, Formalisms, Correctness, 1989.

[12] R. Bagrodia. *An environment for the design and performance analysis of distributed systems*. PhD thesis, The University of Texas at Austin, Austin, Texas, 1987.

[13] L. Bougé and N. Francez. A compositional approach to superimposition. In *ACM Conference on Principles of Programming Langauges*, 1988.

[14] K. Chandy and J. Misra. *Parallel Program Design: A Foundation*. Addison–Wesley, 1988.

[15] E. W. Dijkstra. *A Discipline of Programming*. Prentice–Hall International, 1976.

[16] E. W. Dijkstra, L. Lamport, A. J. Martin, and C. S. Scholten. On-the-fly garbage collection: An exercise in cooperation. *Communications of the ACM*, 21:966 – 975, 1978.

[17] C. A. R. Hoare. Communicating sequential processes. *Communications of the ACM*, 21(8):666–677, August 1978.

[18] INMOS Ltd. *occam -2 Reference Manual.* Prentice–Hall International, 1988.

[19] INMOS Ltd. *The Transputer Reference Manual.* Prentice–Hall International, 1988.

[20] S. Katz. A superimposition control construct for distributed systems. Technical Report STP - 286 - 87, MCC, 1987.

[21] F. Mattern. Algorithms for distributed termination detection. *Distributed Computing*, 2:161 – 175, 1987.

[22] S. Ramesh and S. L. Mehndiratta. A methodology for developing distributed programs. *IEEE Transactions on Software Engineering*, SE–13(8):967–976, 1987.

[23] K. Sere. Stepwise removal of virtual channels in distributed algorithms. In *2nd International Workshop on Distributed Algorithms*, 1987.

# Hoare

Compilation from a high-level language into machine code can be seen as an automated refining of the program to which it is applied. The programmer — the client of the compiler — who is satisfied with the source program expects to be no less satisfied by the resulting object program in machine code. Thus a *correct* compiler is one that produces object programs that do indeed refine their corresponding source programs.

A slight complicating factor however is that source and object programs are usually written in different languages, and so cannot be directly compared. One can deal with that problem by giving both the source and object programming languages a precise meaning within a third language in some mathematical domain. Then the refinement relation is defined there, and a compiler is said to be correct exactly when the meaning of any source program it is given is refined by the meaning of the object code it produces.

In this paper, however, the complication is avoided in a startlingly different way. The meaning of the object programming language is given by providing an interpreter for it that is written in the original source language. Since the mathematical domains mentioned above can be especially intricate, significant simplicity is gained here. The refinement that the compiler should perform is expressible in the source language itself, and the rules for checking that it does so are the same ones that might be used by the high-level programmer anyway, during the development of his source program.

# Refinement Algebra proves correctness of compiling specifications

## C.A.R. Hoare*

### Abstract

A compiler is specified by a description of how each construct of the source language is translated into a sequence of object code instructions. The meaning of the object code can be defined by an interpreter written in the source language itself. A proof that the compiler is correct must show that interpretation of the object code is at least as good (for any relevant purpose) as the corresponding source program.

The proof is conducted using standard techniques of data refinement. All the calculations are based on algebraic laws governing the source language. The theorems are expressed in a form close to a logic program, which may be used as a compiler prototype, or as a check on the results of a particular compilation. It is suggested that this formal framework provides appropriate interfaces for compiler implementors, and hardware designers, as well as users of the language.

## 1 Introduction

Compilation is specified as a relation between a source program $p$ (in some high level source language $H$) and the corresponding object code $c$ (in some target machine language $M$). Further details of compilation are given by a symbol table $t$, mapping the global identifiers of $p$ to storage locations of the target machine. This compilation relation will be abbreviated as a predicate

$$p \subseteq_t c.$$

The internal structure of $p, t$ and $c$ will be elaborated as the need arises.

Improvement is a relation between a product $q$ and a product $p$ that holds whenever for any purpose the observable behaviour of $q$ is as good as or better than that of $p$; more precisely, if $q$ satisfies every specification satisfied by $p$, and maybe more. For example, in a procedural language, a program is better if it terminates more often and/or gives a more determinate result. This relation is written

---

*Programming Research Group, 11 Keble Rd., Oxford OX1 3QD, UK.

$$p \sqsubseteq q,$$

where $\sqsubseteq$ is necessarily transitive and symmetric (a preorder).

If $p$ and $q$ are programs operating on different data spaces, they cannot be directly compared. But if $r$ is a translation from the data space of $q$ to that of $p$, we can then compare them before and after the translation

$$r; p \sqsubseteq q; r,$$

where the semicolon denotes sequential composition. The relation $r$ is known as a simulation or refinement; such simulations are the basis of several modern program development techniques, e.g. VDM, Z.

To define compiler correctness precisely, we need to ascribe meanings to $p, t$, and $c$. Let $\tilde{c}$ be a formal description of the behaviour of the target machine executing the machine code $c$. Let $\tilde{p}$ similarly be an abstract behavioural definition of the meaning of the program $p$. Finally, let $\tilde{t}$ be a transformation which assigns to each global identifier $x$ of the source program the value of the corresponding location $tx$ in the store of the target computer. In order to prove that the object code $c$ is a correct translation of $p$, we need to show that

$$\tilde{t}; \tilde{p} \sqsubseteq \tilde{c}; \tilde{t},$$

or in words, that the source program is improved by execution of its object code. A compiler is correct if it ensures the above for all $p, t, c$. We therefore define this to be the specification of the compiler

$$p \sqsubseteq_t c \ \triangleq \ \tilde{t}; \tilde{p} \sqsubseteq \tilde{c}; \tilde{t}.$$

The task of proof-oriented compiler design is to develop a mathematical theory of the relation $\sqsubseteq$. This should include enough theorems to enable the implementor to select correct object code for each construct of the programming language. It may be that the theory allows choice between several differing object codes for the same source code; this gives some scope for optimisation by selecting the most efficient alternative.

As examples of the kind of theorem by which the designer specifies a complier, consider the following three possible theorems (using $\langle$ and $\rangle$ as sequence brackets and $\frown$ for catenation):

$$(x := y) \sqsubseteq_t \langle \text{load } ty, \text{store } tx \rangle$$

$$\text{skip} \sqsubseteq_t \langle \rangle$$

$$(p; q) \sqsubseteq_t (c1 \frown c2) \text{ whenever } p \sqsubseteq_t c1 \text{ and } q \sqsubseteq_t c2.$$

When these have been proved, the implementor of the compiler knows that a simple assignment can be translated into the pair of commands shown above; that a **skip** command translates to an empty code sequence; and that sequential composition can be translated by concatenating the translations of its two components.

The form of these theorems is extraordinarily similar to that of a recursively defined Boolean function, which could be used to check that a particular compilation of a safety critical program has been successful. It is also quite similar to a logic program, which could be useful if a prototype of the compiler is needed, perhaps for bootstrapping purposes. It is also similar to that of a conventional procedural compiler, structured according to the principle of recursive descent. Finally, it is quite similar to an attribute grammar for the language; and this permits the use of standard techniques for splitting a compiler into an appropriate number of passes. Thus the collection of theorems serves as an appropriate interface between the designers of a compiler and its implementors.

## 2 Interfaces

The correctness condition has three free variables ($p, t$, and $c$), and is comprehensible only in a formal framework in which the following definitions are available.

(1) the semantics of the language $H$, which defines the relation between a program $p$ and its behaviour $\tilde{p}$.

(2) the architecture of the machine, which defines the relationship between machine code $c$ in language $M$ and its behaviour $\tilde{c}$.

(3) the improvement relation $\sqsubseteq$ and the transformation $\tilde{t}$.

Our first task is to choose the most appropriate mathematical framework within which the define the details of H, M, $\sqsubseteq$. In this selection, the most important criterion is that each formal definition must be suitable as the interface to another group of engineers responsible for its implementation and/or use. Furthermore each definition should be self-sufficient for the needs of each class of user, and protect them from the details and complexities of the other interfaces. This means that there will be a significant intellectual gap between the interfaces; and so closing each gap will be a significant mathematical achievement, not a mere recoding exercise. All responsibility for the mutual consistency of the interfaces should be taken by the designers of the high level language and its compiler. Of course, in the actual implementation of the compiler there may be many internal interfaces, for example between successive compilation passes and the loader. Such interfaces are of no concern to any user of the compiler, and will not be treated in this paper.

### 2.1 The high level language

In the design of an implementation of a programming language, the first and absolute requirement is a perfect comprehension of its meaning. If the implementation is to be supported by mathematical proof, the meaning must be expressed by some mathematical definition which forms the basis of the reasoning. A wide variety of formalisms have been proposed for this purpose, and there is difficulty in choosing between them.

The user of a programming language also needs a sufficient understanding of it to discharge an obligation to deliver programs which meet their specification. The specification of a program can be expressed as a mathematical predicate, describing observable characteristics of the behaviour of the program when executed. For example, in the case of a conventional language the specification states the desired relationship between the initial and the final values of the global variables of the program, together with any interactions (input or output) in which it engages during execution. From the users' point of view, the best formalisation of the meaning of a programming language is one that relates programs to the specifications which they satisfy.

The specification-oriented semantics of the programming language formally defines the relation between $p$ and $\tilde{p}$ mentioned in the introduction: $\tilde{p}$ is just the strongest behavioural specification satisfied by $p$. However, in mathematics it is permissible (and indeed universal practice) to use notations directly to stand for what they mean. We can profit enormously by this convention if we just allow the program text itself to mean its strongest specification; thus we define ~ on programs as the identity function:

$$\tilde{p} \;\hat{=}\; p.$$

## 2.2 Machine Language

The whole purpose of designing a compiler is that programmers should not write in machine code. It is the responsibility of the compiler to ensure that execution of the target code should have the same (or better) behaviour than that ascribed to the source code. So it is important that the applications programmer should never need to see the specification of the machine code.

The true user of the formal specification of the target machine code is the hardware engineer, who is responsible for design of the logic of the computer itself, and for certifying that it meets its specification. The formalism that most simplifies this task will be highly operational, including details of the structure of the registers, storage, and input/output devices, and of how they are updated by each instruction. Experience shows that a direct and widely understood way of describing these things is by the notations of a simple procedural programming language.

Apart from familiarity and operational bias, there is another advantage of procedural notations, that of modularity. The definition can start with a simple subset of the components of the machine state, and the subset of instructions which refer to them. New components can later be added, as required to deal with the further instructions which update them. All the earlier simpler program fragments leave the new components unchanged; they remain correct, and their proofs are still valid, and do not have to be changed when moving from the simpler instruction set to the more complex.

But a decisive advantage of defining the target machine code in a high level language can be obtained if the language used for writing the interpreter is a subset of the source language, whose semantics are already known. So the relation between the object code $c$

and its behaviour $\tilde{c}$ is that $\tilde{c}$ is a program written in a high level language. Its initial state is known to contain the object code $c$ produced by the compiler, and its task is to interpret that code, engaging in the input and output instructions and producing eventually the final machine state. It is the task of the hardware designer to implement a fast interpreter by use of a different technology.

## 2.3 The improvement relation

The improvement relation is now one that must be defined between two programs $p$ and $\tilde{c}$; both of them are expressed in the same high level programming language, which has a known specification-oriented semantics. The only possible definition of improvement must be in terms of specifications: the better program must satisfy every single specification satisfied by the worse program, and maybe more. Thus whenever the worse program $p$ is known to be adequate for a particular task, the better one $q$ will be too. The usual notation for this is

$$p \sqsubseteq q.$$

This kind of direct comparison is only possible between programs operating on the same global variables, and specifications expressed in terms of them. But our interpreter $\tilde{c}$ operates on variables denoting the registers and store of the target computer, whereas the source program $p$ operates on variable identifiers usually chosen by the programmer. In order to compare them, we need to know the relationship between the programmer's variable names and the machine locations to which they have been allocated by the compiler. This is defined by the compiler's symbol table $t$, which maps each variable of the program $p$ to the address of the storage location allocated to hold its value. We therefore define $\tilde{t}$ as a sort of symbolic dump; it assigns to each high-level variable $x$ the value of its corresponding storage location, thereby effecting a transformation from the low level machine state to the implicit state of the high level language program. Thus $\tilde{t}$ can be written as an assignment

$$x := M[tx];\ y := M[ty];\ \ldots$$

where $x, y \ldots$ are program variables and $M$ is an array representing the variable store of the target machine. Now we can define the concept of improvement as one that holds between programs when their corresponding states have been transformed by $\tilde{t}$:

$$\tilde{t};\ p \sqsubseteq \tilde{c};\ \tilde{t}.$$

On the left hand end of this formula, $\tilde{t}$ initialises the program variables from the target store; and on the right hand end it extracts the final values of the variables after execution of the object code. It may seem strange that the compiler itself is allowed to determine $t$, which plays such an important role in the proof of its own correctness. For example, is there a danger that $\tilde{t}$ might be the program *abort*, which would make the compilation relation trivially true? Fortunately, the construction of $\tilde{t}$ avoids these and other dangers.

It is one of the goals of this paper to make a convincing case that theorems being proved are actually relevant and conducive to the correctness of compilers.

## 2.4 Proof Strategy

The compilation predicate is defined as an inequation. The easiest way to prove such inequations is by algebraic transformations and simplifications. Such transformations can readily be checked or even performed by computer. But each transformation must be justified by an algebraic law which is valid for the language in which the reasoning is conducted. So these algebraic laws must themselves be proved from the specification-oriented semantics of the programming language.

Thus it seems a good idea to separate from the task of proving a particular compiler the more general task of deriving a sufficient set of algebraic laws applicable to all programs expressed in the language. These laws are of especial value in optimisation of the source or target codes. The sufficiency of such a set of algebraic laws can be established by an appropriate kind of normal form theorem. In effect, a complete set of laws can act as an algebraic specification of the meaning of the high-level programming language; and one which is far more suited to the needs of compiler design than the specification-oriented semantics from which they were derived. Another advantage of algebraic laws is that of modularity and generality — each of them is valid in many different programming languages.

The intellectual gap between a specification-oriented semantics and an algebraic semantics is a significant one, and well worth treating as a separate task, which should ideally be carried out by the designers of the programming language. In the remainder of this paper, we will assume that the task has already been completed. In the next section we state without proof the algebraic laws which we will assume to be true of our chosen source language, and state some useful general theorems that can be proved from them. The following section will show how the theorems can be used to prove parts of a simple compiler.

# 3 Algebraic semantics

The basic laws defining Dijkstra's language [1] of guarded commands are given in [2]. Some of the more useful ones are repeated here for convenience.

(1) Sequential composition is associative and has unit **skip**.

(2) Conditionals are coproducts:

$$(\text{if true then } p \text{ else } q) = p = (\text{if false then } q \text{ else } p)$$
$$(\text{if } b \text{ then } p \text{ else } q); r = \text{if } b \text{ then } (p; r) \text{ else } (q; r).$$

(3) If $e$ does not contain $x$, assignment obeys substitution laws like

$$(x := e; y := e) = (y := e; x := y)$$
$$(x := f; x := e) = x := e$$
$$x := e; \text{if } x \text{ then } p \text{ else } q$$
$$= \text{if } e \text{ then } (x := e; p) \text{ else } (x := e; q).$$

(4) $\sqsubseteq$ is an $\omega$-complete partial order, i.e., it is reflexive, transitive and antisymmetric; and if $p_i \sqsubseteq p_{i+1}$ for all $i$, then the sequence has a least upper bound satisfying

$$\sqcup_i p_i \sqsubseteq q \text{ iff (for all } i: \ p_i \sqsubseteq q).$$

(5) **abort** is the bottom of $\sqsubseteq$ and left zero of sequential composition. Further, if $p$ contains no input or output

$$p; \textbf{abort} \ = \ \textbf{abort}.$$

(6) Least fixed points are defined for a program $\mathcal{P}X$ containing the program variable $X$:

$$\mu X.\mathcal{P}X = \sqcup_i \mathcal{P}^i(\textbf{abort}).$$

(7) The **while** is defined

$$\textbf{while } b \textbf{ do } p$$
$$= \mu X. \textbf{ if } b \textbf{ then } (p;\ X) \textbf{ else skip}.$$

(8) $\sqsubseteq$ has a greatest lower bound operator $\sqcap$, representing non-deterministic choice:

$$(r \sqsubseteq p \text{ and } r \sqsubseteq q) \text{ iff } r \sqsubseteq (p \sqcap q).$$

From this it follows that

$$(p \sqcap q) \sqsubseteq p \text{ and } p \sqcap q = q \sqcap p.$$

(9) Declaration introduces a new variable. If $e$ always terminates then

$$(\textbf{var } v;\ v := e) \ = \ \textbf{skip},$$

and initialisation of a variable can only make a program more deterministic:

$$(\textbf{var } v;\ p) \sqsubseteq (\textbf{var } v;\ v := e;\ p).$$

If $p$ contains no occurrence of $v$ then

$$p;\ (\textbf{var } v;\ q) \ = \ (\textbf{var } v;\ p;\ q).$$

The usual laws for substitution of bound variables are valid.

(10) We define an assertion as causing abortion if false

$$\{b\} \ \triangleq \ \textbf{if } b \textbf{ then skip else abort}.$$

It follows that

$$\{b\} \sqsubseteq \text{skip}.$$

If $e$ does not contain $x$ then

$$x := e = (x := e; \{x = e\})$$
$$\text{and} \quad (\{x = e\}; x := e) = \{x = e\}.$$

From these and other basic laws, reported in [2], a number of useful theorems can be derived:

- If $b \Rightarrow c$ then (while $b$ do $p$; while $c$ do $p$) $\sqsubseteq$ while $c$ do $p$.
- If $\mathcal{P}Y \sqsubseteq Y$ then $\mu X.\mathcal{P}X \sqsubseteq Y$.
- If (for all $X, Y : X; p \sqsubseteq p; Y \Rightarrow \mathcal{P}X; p \sqsubseteq p; \mathcal{Q}Y$) then $\mu X.\mathcal{P}X; p \sqsubseteq p; \mu Y.\mathcal{Q}Y$.

The last of the above is used to lift simulation through recursion. (See for example theorem 4.)

# 4 The interpreter

In our simple example, we shall treat a machine with just four components.

$$m : \text{rom} \to \text{instruction}$$

is the store occupied by the code of the program;

$$M : \text{ram} \to \text{word}$$

is the store used for variables (where the word-length is unspecified);

$$P : \text{rom}$$

is the pointer to the current instruction, and

$$A : \text{word}$$

is the only general-purpose register.

We will consider only four instructions, each defined by a fragment of code describing its effect for all $n$:ram and $j$:rom.

$$\begin{aligned}
\text{effect (load } n) &= (A := M[n]; P := P+1) \\
\text{effect (store } n) &= (M[n] := A; P := P+1) \\
\text{effect (jump } j) &= P := j \\
\text{effect (cond } j) &= \text{if } A \text{ then } P := P+1 \text{ else } P := j.
\end{aligned}$$

For all undefined instructions $i$, we define

$$\text{effect } (i) = \text{abort}.$$

This places upon the compiler designer the obligation to ensure that such an instruction is never executed. More importantly, it leaves open the option of later definition of additional instructions. All earlier proofs remain valid, because the interpreter can only be improved by making it abort less often.

We have now defined the effect of each single machine instruction. But an interpreter is required to interpret a whole sequence of instructions held in $m$ between two specified locations, say $s$ and $f$. The interpreter $Isf$ accomplishes this by means of a loop

$$\begin{aligned}
Isf \;\hat{=}\;\; & P := s; \\
& \text{while } P < f \text{ do } step; \\
& \{P = f\},
\end{aligned}$$

where $step$ abbreviates effect $(m[P])$.

The final assertion $\{P = f\}$ places upon the compiler designer an obligation to compile code which exits normally through its end, without any wild forward jumps. This is essential to prove correctness of sequential composition. A hardware implementation of the interpreter is permitted to ignore the assertions; that can only make it better than the one which is known to be correct.

Elementary facts about the interpreter can be derived by symbolic execution in the case that $s < f$. For example

**Lemma 0.**

$$\begin{aligned}
&\text{If} \quad && m[s] &&= (\text{load } n) \text{ and } s < f \\
&\text{then} \quad && Isf &&= A := M[n]; I(s+1)f.
\end{aligned}$$

$$\begin{aligned}
&\text{If} \quad && m[s] &&= (\text{store } n) \text{ and } s < f \\
&\text{then} \quad && Isf &&= M[n] = A : I(s+1)f.
\end{aligned}$$

$$\begin{aligned}
&\text{If} \quad && m[s] &&= (\text{load } l) \text{ and } m[s+1] = (\text{store } n) \text{ and } s+1 < f. \\
&\text{then} \quad && Isf &&= (A := M[n]; M[l] := A; I(s+2)f)
\end{aligned}$$

$$\begin{aligned}
&\text{If} \quad && m[s] &&= (\text{jump } j) \text{ and } s < f \\
&\text{then} \quad && Isf &&= Ijf.
\end{aligned}$$

We can already prove a series of important lemmas.

**Lemma 1.** *Iff* = $(P := f)$.

Proof:

$$
\begin{aligned}
&\quad\text{LHS} \\
&= P := f;\ \text{while } P < f \text{ do} \ldots;\ \{P = f\} \\
&= P := f;\ \text{if } f < f \text{ then} \ldots \text{else skip};\ \{P = f\} \\
&= P := f.
\end{aligned}
$$

□

**Lemma 2.** If $s < f$ and $m[s] = (\text{jump } f)$ then $Isf = (P := f)$.

Proof: Lemmas 0 and 1. □

**Lemma 3.** If $j \leq f$ then $Isj;\ Ijf \sqsubseteq Isf$.

Proof:

$$
\begin{aligned}
&\quad\quad\quad\quad\quad\quad\quad\text{LHS} \\
\{\text{by definition}\} \;=\;& P := s;\ \text{while } P < j \text{ do } step;\ \{P = j\}; \\
& P := j;\ \text{while } P < f \text{ do } step;\ \{P = f\}
\end{aligned}
$$

$\{\text{omit } P := j \text{ since } P = j,\text{ then omit assertion too}\}$
$$
\begin{aligned}
\sqsubseteq\;& P := s;\ \text{while } P < j \text{ do } step; \\
& \text{while } P < f \text{ do } step;\ \{P = f\}
\end{aligned}
$$

$\{\text{since } j \leq f \wedge P < j \Rightarrow P < f, \text{ and loop with stronger guard can be omitted}\}$
$$
\begin{aligned}
\sqsubseteq\;& P := s;\ \text{while } P < f \text{ do } step;\ \{P = f\} \\
\{\text{by definition}\} \;=\;& Isf.
\end{aligned}
$$

□

**Lemma 4.** If $s < j \leq f$ and $m[s] = (\text{cond } j)$ then $Isf = (\text{if } A \text{ then } I(s+1)f \text{ else } Ijf)$.

Proof:

$\quad Isf$

$\{\text{expansion, definition of effect } (\text{cond } j)\}$
$$
\begin{aligned}
=\;& P := s;\ (\text{if } A \text{ then } P := P + 1 \text{ else } P := j); \\
& \text{while } P < f \text{ do } step;\ \{P = f\}
\end{aligned}
$$

$\{\text{distribution, overriding}\}$
$$
\begin{aligned}
=\;& \text{if } A \text{ then } (P := s + 1;\ \text{while} \ldots;\ \{P = f\}) \\
& \text{else } (P := j;\ \text{while} \ldots;\ \{P = f\})
\end{aligned}
$$

$\{\text{definition of } I\}$
$\quad = \text{RHS}.$

□

**Lemma 5.** If $s + 1 < j$ and $m[j-1] = (\text{jump } f)$ are added to the conditions of lemma 4 then

$$(\text{if } A \text{ then } (I(s+1)(j-1); P := f) \text{ else } Ijf) \sqsubseteq Isf.$$

Proof:

$$\begin{array}{rl} & I(s+1)f \\ \{\text{lemma 3}\} \sqsupseteq & I(s+1)(j-1); I(j-1)f \\ \{\text{lemma 2}\} = & I(s+1)(j-1); P := f. \end{array}$$

The conclusion follows from that of lemma 4 by monotonicity. □

## 5  The compiler specification

The compiling predicate $p \underset{t}{\sqsubseteq} c$ means

$$\tilde{t}; p \sqsubseteq \tilde{c}; \tilde{t}.$$

In our simple example, the object code $c$ is specified as a triple $(m, s, f)$, where $m$ is the code store, $s$ is the starting address and $f$ is the finishing address for execution. $\tilde{c}$ is then defined as $Isf$. The symbol table is a finite injection

$$t : \text{identifier} \rightarrowtail \text{ram}.$$

It is used to define the simulation $\tilde{t}$ as a multiple assignment to all the non-local variables of the program

$$\tilde{t} \;\hat{=}\; (x, y, z, \ldots := M[tx], M[ty], M[tz], \ldots).$$

This also forgets the machine state held in the variables $P, A, M$.

We can now begin to prove the series of theorems by which a compiler designer specifies to its implementor the valid object code for each of the constructions of the source language.

**Theorem 1** (skip).

$$\text{skip} \subseteq_t (m, f, f).$$

Proof:

$$\begin{array}{rl} & \textit{Iff}; \tilde{t} \\ \{\text{lemma 1}\} = & (P := f; \tilde{t}) \\ \{\tilde{t} \text{ forgets } P\} = & \tilde{t} \\ \{\text{identity}\} = & \tilde{t}; \text{skip}. \end{array}$$

□

There is no need to specify an unique object code for each source program construction. For example, lemma 2 also permits us to prove

**Theorem 1a.** If $s < f \wedge m[s] = (\text{jump } f)$ then $\text{skip} \subseteq_t (m, s, f)$. □

Sequential composition is the subject of a more complicated theorem; it specifies that a sequential composition can be compiled by recursively compiling its operands.

**Theorem 2** (composition).

If $\quad s \leq j \leq f$ and $p \subseteq_t (m, s, j)$ and $q \subseteq_t (m, j, f)$
then $\quad (p; q) \subseteq_t (m, s, f).$

Proof: From the antecedent we have

$$\begin{array}{rl} & \tilde{t}; p; q \\ \{\text{monotonicity}\} \sqsubseteq & Isj; \tilde{t}; q \\ \{\text{monotonicity}\} \sqsubseteq & Isj; Ijf; \tilde{t} \\ \{\text{lemma 3}\} \sqsubseteq & Isf; \tilde{t}. \end{array}$$

□

**Theorem 3** (conditional).

If $\quad s + 2 < j < f$
and $\quad m[s] = (\text{load } tb)$ and $m[s + 1] = (\text{cond } j)$
and $\quad m[j - 1] = (\text{jump } f)$
and $\quad p \subseteq_t (m, s + 2, j - 1)$ and $q \subseteq_t (m, j, f)$
then $\quad (\text{if } b \text{ then } p \text{ else } q) \subseteq_t (m, s, f).$

Proof:

$$Isf; \tilde{t}$$

{lemmas 5 and 0}
$\sqsupseteq \quad A := M[tb];$
    (if $A$ then $(I(s+2)(j-1); P := j)$ else $Ijf); \tilde{t}$

{distribution, using antecedents, and forgetting $P$}
$\sqsupseteq \quad A := M[tb];$
    if $M[tb]$ then $(\tilde{t}; p)$ else $(\tilde{t}; q)$

{distribution through conditional, and forgetting $A$}
$= \quad \tilde{t};$ if $b$ then $p$ else $q$.

□

**Theorem 4** (while).

If $s+1 < f-1$
and $m[s] = $ (load $tb$) and $m[s+1] = $ (cond $f$)
and $m[f-1] = $ (jump $s$)
and $p \subseteq_t (m, s+2, f-1)$
then (while $b$ do $p$) $\subseteq_t (m, s, f)$.

Proof:

$\phantom{\{lemma 3\}} \quad I(s+2)f$
{lemma 3} $\quad \sqsupseteq \quad I(s+2)(f-1); I(f-1)f$
{$m[f-1]$} $\quad = \quad I(s+2)(f-1);$
$\phantom{\{lemma 3\} \sqsupseteq} \quad P := s;$ while $P < f$ do step; $\{P = f\}$
{definition $I$} $\quad = \quad I(s+2)(f-1); Isf.$

Now by symbolic execution,

$\phantom{\{above\}} \quad Isf$
$\phantom{\{above\}} = \quad A := M[Tb];$
$\phantom{\{above\} = \quad} $ if $A$ then $I(s+2)f$ else $P := f$
{above, and monotonicity} $\sqsupseteq \quad A := M[Tb];$
$\phantom{\{above, and monotonicity\} \sqsupseteq} $ if $A$ then $I(s+2)(f-1); Isf$ else $P := f$.

Thus if we define

$\mathcal{P}X \quad \hat{=} \quad A := M[tb];$
$\phantom{\mathcal{P}X \quad \hat{=} \quad} $ if $A$ then $(I(s+2)(f-1); X)$ else $P := f$,

we have shown that $\mathit{Isf} \sqsupseteq \mathcal{P}(\mathit{Isf})$ and hence that

$$\mathit{Isf} \sqsupseteq \mu X.\mathcal{P}X. \tag{1}$$

Now assume $X; \tilde{t} \sqsupseteq \tilde{t}; Y$. Then, as in the proof of theorem 3,

$$\mathcal{P}X; \tilde{t} \sqsupseteq \tilde{t}; \text{ if } b \text{ then } (p; Y) \text{ else skip}.$$

By lifting the simulation through recursion

$$\begin{aligned}(\mu X.\mathcal{P}X); \tilde{t} & \sqsupseteq \tilde{t}; \mu Y. \text{ if } b \text{ then } (p; Y) \text{ else skip} \\ \{\text{definition while}\} & = \tilde{t}; \text{ while } b \text{ do } p.\end{aligned}$$

The result now follows from (1). □

**Theorem 5** (non-determinism).

If $p \underset{t}{\sqsubseteq} (m,s,f)$ then $(p \sqcap q) \underset{t}{\sqsubseteq} (m,s,f)$.

Proof:

$$\begin{aligned}& \tilde{t}; (p \sqcap q) \sqsubseteq \tilde{t}; p \\ \{\text{by assumption}\} \sqsubseteq\ & \mathit{Isf}; \tilde{t}.\end{aligned}$$

□

Since $p \sqcap q = q \sqcap p$, we have immediately

**Theorem 5a.** If $q \underset{t}{\sqsubseteq} (m,s,f)$ then $(p \sqcap q) \underset{t}{\sqsubseteq} (m,s,f)$. □

This gives the implementor freedom to implement $p \sqcap q$ either as $p$ or as $q$. This freedom may be used to improve the quality of the compiler by selecting whichever gives the more efficient object code, at least in the cases where this is statically checkable.

**Theorem 6** (abort).

$$\text{abort} \underset{t}{\sqsubseteq} (m,s,f).$$

Proof: Since $\tilde{t}$ contains no input or output

$$\tilde{t}; \text{abort} = \text{abort} \sqsubseteq \mathit{Isf}; \tilde{t}.$$

□

If the program would abort, the compiler is allowed to generate any object code whatsoever.

The theorem for declaration applies only when the declared variable is distinct from all global variables. (In the absence of recursion, this can be achieved by preliminary substitution of a fresh name for the local variable). In any case, the body of the block is compiled with a symbol table $t'$, an injection formed by extending the domain of $t$ to include the declared variable $v$. Thus $t$ equals $v \triangleleft t'$, i.e., the result of removing $v$ from the domain of $t'$.

**Theorem 7** (declaration).

If $t = v \triangleleft t'$ and $p \subseteq_{t'} (m, s, f)$
then $(\text{var } v; p) \subseteq_{t} (m, s, f)$.

Proof:

$$
\begin{array}{rl}
& \tilde{t}; (\text{var } v; p) \\
\{v \text{ fresh}\} = & (\text{var } v; \tilde{t}; p) \\
\{\text{law}\} \sqsubseteq & (\text{var } v; v := M[t'v]; \tilde{t}; p) \\
\{\text{def } \sim\} = & (\text{var } v; \tilde{t'}; p) \\
\{\text{antecedent}\} \sqsubseteq & (\text{var } v; \mathit{Isf}; \tilde{t'}) \\
= & \mathit{Isf}; (\text{var } v; \tilde{t'}) \\
= & \mathit{Isf}; \tilde{t}; (\text{var } v; v := M[t'v]) \\
= & \mathit{Isf}; \tilde{t}.
\end{array}
$$

□

# 6 Conclusion

This paper has illustrated a structure and methodology for proof of correctness of compiler designs; the proofs use primarily algebraic transformations to establish refinement relations. But the example programming language is very small, and there are at present no grounds to suppose that the methods will generalise to more powerful source languages or more complex target machines.

One hopeful aspect of the method is that each feature of the language is introduced by a separate theorem. So new features should be easy to add, as well as new, more efficient, translations of existing features.

On the other hand, it is clear that major new features such as recursion or concurrency (or worse still, both of them) will require significant additional complexity, for example

(1) inclusion of new components such as stacks into the machine state;

(2) formulation of invariant assertions (e.g. non-interference and stack integrity) and their insertion into the interpreter;

(3) proliferation of compiler tables to deal with scopes and types.

Ideally, these additions should preserve the validity of all the proofs about the simpler aspects of the language.

## Acknowledgement

The research reported here is a contribution to an international project PROCOS (Provably correct systems), supported by the European Community under ESPRIT II Basic Research Actions. It owes much to the inspiration and assistance of many colleagues on that project, and to detailed scrutiny by Carroll Morgan.

## References

[1] E.W. Dijkstra. Guarded commands, non-determinacy, and formal derivation of programs. Comm. ACM 18 8 (1975) 453–7.

[2] Laws of Programming. C.A.R. Hoare, J.M. Spivey and others. Comm. ACM 30 8 (1987) 672-86.

# Jones

One cannot escape the fact that in some programming languages there are programs that do not terminate cleanly. Although for any expression $E$ in ordinary logic the formula $E = E$ is trivially true, one cannot assume for example that the Boolean expression (1/0).EQ.(1/0) therefore evaluates to .TRUE. — more likely is that is doesn't evaluate at all.

The stark disagreement between logic and the behaviour of some programs must be reconciled if a refinement method is to be shown — as it must be — to be correctness preserving. It is no good having grand plans for specifications and their refinement methods if there is a fundamental mismatch at the lowest level. Thus one must change either the programming language or the logic.

An attractive approach therefore is to devise a logic that contains, like the programming language, expressions that 'do not terminate'; they are the *partial functions*, so-called to distinguish them from total functions. The function taking real-valued $x$ to the value $x + 1$ is total, because it is defined for all $x$. But the function $1/x$ is partial — it is not defined when $x$ is 0. A logic for partial functions can thus be used to ensure the soundness of a refinement method that leads to programs containing non-terminating expressions, as most do. The following paper describes such a logic.

# On the usability of logics which handle partial functions

J.H. Cheng[*], C.B. Jones[†]

**Abstract**

Partial functions are common in computing – they arise from recursive definitions and from the basic operators of data types like sequences. The need to prove results about logical formulae which include terms using partial functions must cast doubt on the applicability of classical logic. Approaches to handling partial functions which attempt to salvage classical logic include avoiding function application and the use of variant equality notions. The alternative is to accept the need for a non-classical logic. The most frequently used non-classical logic adopts non-strict (but also non-commutative) conditional versions of the propositional operators. An alternative – which is more fully explored below – uses commutative and/or operators as defined in the truth-tables which Kleene attributes to Łukasiewicz. A consistent and complete proof system for such a logic has been used in the program development method known as VDM. An important feature of this approach is the way the definitions of partial functions are brought into proofs. This paper analyzes different approaches to reasoning about partial functions. In particular, it compares the usability of different logics in handling the sort of partial functions which arise in the specification and design of computer software.

## 1 Introduction

A partial function is one whose application does not necessarily yield a (defined) value. Partial functions are often defined recursively; for example:

$zero : Z \to Z$

$zero(i) \triangleq$ if $i = 0$ then $i$ else $zero(i-1)$

defines a function for which the property:

$\forall i \in Z \cdot i \geq 0 \implies zero(i) = 0$

might be expected to hold ($Z$ is the set of integers). Closer investigation discloses, however, that this logical expression is problematic with the classical logical operators and notion of equality: the bound on $i$ permits a value of $-1$ for which, although the antecedent of the implication is false, the consequent contains a term ($zero(-1)$) which – under reasonable interpretations of the definition of $zero$ – does not denote an integer. In order to make this plausible quantified

---

[*]I.S.T., Cambridge
[†]Department of Computer Science, University of Manchester

expression have a meaning, some way of handling terms which include such references to partial functions must be found. In this simple example, the problem could be circumvented by changing the bound on the quantifier to range over the natural numbers ($\mathbb{N}$). Another alternative is to define equality so as to prevent the undefinedness of the term causing an undefined logical expression. In [BCJ84], a logic is discussed in which non-strict logical operators handle the difficulty.

This paper analyzes different approaches to the construction of formal proofs about partial functions. The aim is to discuss alternative approaches and to compare some of the problems which arise in their application to various examples. The body of the paper (Sections 2–9) presents the main approaches found in the literature; Section 10 discusses a number of related papers; and the conclusion (Section 11) indicates why the current authors believe their approach to be usable in software design.

Before embarking on the analysis of different approaches, it is worth taking a closer look at the equality relation. In particular, recognising that different notions of equality are appropriate in different contexts is the key to some approaches to handling partial functions. (One might want to avoid manipulating several forms of equality in proofs about functions, but the underlying distinction should first be understood.) It should be clear that the = within the conditional expression must be computable and therefore strict (i.e. yields $\bot$ if either operand is $\bot$). Writing the undefined values as 'bottom' elements ($\bot$), its truth-table is:

| =         | 0     | 1     | 2     | ... | $\bot_\mathbb{N}$ |
|-----------|-------|-------|-------|-----|-------------------|
| 0         | true  | false | false |     | $\bot_B$          |
| 1         | false | true  | false |     | $\bot_B$          |
| 2         | false | false | true  |     | $\bot_B$          |
| ⋮         |       |       |       |     |                   |
| $\bot_\mathbb{N}$ | $\bot_B$ | $\bot_B$ | $\bot_B$ |     | $\bot_B$          |

The definitional equality needs a different interpretation. Operationally speaking, applying the definition on *zero* to $-1$ rewrites to exactly *zero*($-2$): they are strongly equal in the sense that the former is $\bot$ precisely because the latter is $\bot$. The truth-table for this non-strict operator is:

| ==        | 0     | 1     | 2     | ... | $\bot_\mathbb{N}$ |
|-----------|-------|-------|-------|-----|-------------------|
| 0         | true  | false | false |     | false             |
| 1         | false | true  | false |     | false             |
| 2         | false | false | true  |     | false             |
| ⋮         |       |       |       |     |                   |
| $\bot_\mathbb{N}$ | false | false | false |     | true              |

It is, of course, now standard to view recursive definitions like that for *zero* as denoting the least-fixed point of their defining equation (see [Man74, Sto77, Sch86]). But, since the least-fixed point in this case is $\{(i, 0) \mid i \in \mathbb{N}\}$, this does nothing to resolve the problem of the non-denoting term.

Apart from recursively defined functions, many basic functions (which are often written as prefix or infix operators) in software specification and design are also partial. For example, in the notation of [Jon90], writing hd for the head of a sequence, tl for its tail, and [ ] for an empty sequence:

$\forall t \in \mathbb{N}^* \cdot t = [\,] \vee t = \text{append}(\text{hd}\, t, \text{tl}\, t)$

has non-denoting terms in the second disjunct if the first is true; or, with $pred: X \times X \to B$: an expression like:

$$\operatorname{len} t_1 = \operatorname{len} t_2 \wedge \forall i \in \operatorname{dom} t_1 \cdot pred(t_1(i), t_2(i))$$

can index outside the sequence $t_2$ if it is shorter than $t_1$ (and, thus, the first conjunct is false); or again, with $\rho \in (Id \xrightarrow{m} den)$:

$$id \in \operatorname{dom} \rho \wedge \rho(id) \in Proctype$$

the map application can be undefined if the first conjunct is false. (Such expressions occur often, for example, in [BJ82].)

Here, it can be seen that, if the partial terms yield $\perp$, the strict logical operators in disjunctions and conjunctions are inadequate. In the last example, it is also clear that re-interpreting equality does not help since $\in$ is the operator whose first operand is a potentially undefined term.

The following sections analyze various approaches to such examples. Section 6 introduces further problems concerned with potentially undefined predicates being used as the hypotheses of sequents ($\Gamma \vdash p$, where $\Gamma$ denotes a list of formulae which is the hypothesis of the sequent). The approaches in Sections 2, 3, and 4 manage, in various ways, to live with the standard Predicate Calculus; those in Sections 5, 6, and 7 adopt non-standard logics of one kind or another; the approaches in Sections 8 and 9 are influenced by Scott's domain theory, but the former uses a layered-logic approach whereas the latter uses a constructive free logic.

## 2  Totality over a restricted set

As is hinted in the introduction, it is sometimes possible to find a set over which the problem of partiality disappears; thus:

$$\forall n \in N \cdot zero(n) = 0$$

But for other functions, such as:

$$subp : Z \times Z \to Z$$

$$subp(i,j) \triangleq \text{ if } i = j \text{ then } 0 \text{ else } subp(i, j+1) + 1$$

reformulation of properties like:

$$\forall i, j \in Z \cdot i \geq j \Rightarrow subp(i,j) = i - j$$

requires cumbersome expressions such as:

$$\forall i, j \in \{(i,j) \in Z \times Z \mid i \geq j\} \cdot subp(i,j) = i - j$$

(One of the current authors attempted to carry through such a programme using bounded quantifiers in [Jon80].) Another difficulty is that the use of bounds which are given by such general set comprehensions makes the type structure undecidable. As discussed in [Lin87], this results in type inferences like (writing $(X \mid p)$ for the type whose elements are in $\{x : X \mid p(x)\}$):

$$\frac{x \in X;\ p(x)}{x \in (X \mid p)}$$

which could introduce many extra steps into proofs. This problem can be reduced if references to partial functions occur in carefully guarded contexts. Thus, if conditionals are used, inference rules like:

$$\text{if-}I \ \frac{p \vdash e_1 \in Y; \ \neg p \vdash e_2 \in Y}{(\text{if } p \text{ then } e_1 \text{ else } e_2) \in Y}$$

can simplify proofs.

Clearly, the sequence property can be written in terms of a *Nonemptylist* set:[1]

$$\forall t \in \textit{Nonemptylist} \cdot t = \textit{append}(\text{hd } t, \text{tl } t)$$

It is, however, less clear how to translate the property on mappings using bounded quantifiers. In [Jon80] conditional expressions were used as an additional (non-strict) logical operator permitting:

if $id \in \text{dom } \rho$ then $\rho(id) \in \textit{Proctype}$ else false

So bounded quantifiers can be used to avoid some of the problems of partial functions in logical expressions: one must ensure by the bound sets that functions are never applied outside their domains. If suitable constraints could always be given, this would ensure that only the classical (strict) logical operators would be required. These constraints cannot, unfortunately, always be found and it is necessary to introduce at least one non-strict operator (e.g. conditional expressions). It would thus be necessary to develop a proof theory which linked the classical operators and quantifiers to the conditional expressions. Perhaps most worrying is the observation that the attempt to formulate logical expressions with such bounds causes gratuitous difficulty in proof construction. These observations have led a number of researchers to move to the approach – discussed in Section 5 below – in which the conditional concept is absorbed into the meaning of the logical operators.

## 3 Viewing functions as relations

Another way of preserving the strict (classical) propositional operators is to avoid function application! Publications on the specification language known as 'Z' [Abr84b, Hay87, Spi88] appear to use more than one way of handling partial functions. At least [Abr87] proposes avoiding application in the case of recursively defined functions. Instead of writing $f(x) = y$, the function can be treated as its graph and the proposition formulated by testing whether the pair $(x, y)$ is a member of that graph. Thus, the property about *zero* can be formulated:

$$\forall i \in \mathbf{Z} \cdot i \geq 0 \ \Rightarrow \ (i, 0) \in \textit{zero}$$

Since $(-1, j) \in \textit{zero}$ is false for any $j$, the classical (two-valued) interpretation of implication is adequate.

In constructing proofs about recursive functions, an obvious way to treat their definitions is to view them as inducing inference rules:

$$\textit{zero-b} \ \frac{}{(0, 0) \in \textit{zero}}$$

---

[1] It should be noted that it is no longer possible to treat bounded quantification as an abbreviation ($\forall x \in X \cdot p(x)$ as $\forall x \cdot x \in X \ \Rightarrow \ p(x)$, and $\exists x \in X \cdot p(X)$ as $\exists x \cdot x \in X \wedge p(x)$) but it is possible to define inference rules for bounded quantifier introduction which do not rely on these abbreviations. (In fact, this has been done for the Logic of Partial Functions discussed in Section 6.)

$$\text{zero-i} \frac{i \neq 0; (i-1, k) \in \text{zero}}{(i, k) \in \text{zero}}$$

In order to express the property for *subp*, one has to either Curry the function or treat it as a function whose domain is an element of a Cartesian product:

$$\forall i, j \in Z \cdot i \geq j \implies ((i,j), i-j) \in \text{subp}$$

It is unclear how easily one could train oneself to find such expressions as convenient as function applications. Notice that it is also necessary to decide which applications are to be formulated in terms of membership of the graph of the function (all? any, potentially, partial? recursive only?). Furthermore, the reformulation of expressions which are more complex than $f(x) = y$ often needs extra quantifiers: for $p(id) \in \text{Proctype}$, it is necessary to write $\exists y \cdot (x, y) \in p(id) \land y \in \text{Proctype}$ (and it is an extra property of function graphs that avoids the need to write this as $\forall y \cdot (x, y) \in p(id) \implies y \in \text{Proctype}$).

## 4 Fixing notions of equality

The virtue of the proposals in the preceding sections is that they stay within the framework of classical (two-valued) logic; their disadvantage is that they require expressions and proofs involving partial functions to be written in ways which differ significantly from those which are familiar for total functions. This section presents another way of preserving the classical logic operators.

As can be seen from the examples given above, the potentially undefined terms often occur as operands of an equality relation. Can one use different notions of equality? If it can be arranged that the equality operator always yields a Boolean value, the classical logical operators can be used. Such an equality operator can obviously not be strict. The most obvious choice is to use the strong equality discussed in Section 1. (This approach is used in [Bro87b].) Thus, the property to be proved about *subp* is written:

$$\forall i, j \in Z \cdot i \geq j \implies \text{subp}(i, j) == i - j$$

The definition of *subp* is treated as being defined by the same equality:

$$\text{subp}(i, j) == \text{if } i = j \text{ then } 0 \text{ else } \text{subp}(i, j+1) + 1$$

and the following two inference rules are used for reasoning about conditional expressions.

$$\text{if-}E1 \frac{p}{(\text{if } p \text{ then } e_1 \text{ else } e_2) == e_1}$$

$$\text{if-}E2 \frac{\neg p}{(\text{if } p \text{ then } e_1 \text{ else } e_2) == e_2}$$

The proof of the required property is presented in Figure 1 in the style of [Jon90] and uses only classical inference rules for the predicate calculus.

While it is true that the overall structure of this proof is similar to that which might be used for total functions, there do remain some difficulties for the authors of such proofs. The principal difficulty is that both writers and readers of such proofs are forced to be aware of two notions of equality. The (computable) weak equality needed within the definition of a function must not be confused with the strong equality needed to reason about functions. This burden is unacceptable

from $i, j \in \mathbb{Z}$
1     $i = i - 0$     h, $\mathbb{Z}$
2     (if $i = i - 0$ then 0 else $subp(i, (i-0)+1)+1) == 0$     if-$E1(1)$
3     $subp(i, i-0) ==$ if $i = i - 0$ then 0 else $subp(i, (i-0)+1)+1$     $subp$-$defn$
4     $subp(i, i-0) == 0$     $==$-$subs(2,3)$
5     from $n \in \mathbb{N}$; $subp(i, i-n) == n$
5.1       $i - (n+1) \in \mathbb{Z}$     h, h5, $\mathbb{Z}$
5.2       $i \neq i - (n+1)$     h, h5, $\mathbb{Z}$
5.3       (if $i = i - (n+1)$ then 0 else $subp(i, i - (n+1)+1)+1)$     if-$E2(5.2)$
         $== subp(i, i - (n+1)+1)+1$
5.4       $== subp(i, i-n)+1$     $\mathbb{Z}$, 5.3
5.5       $== n+1$     $==$-$subs(h5, 5.4)$
5.6       $subp(i, i-(n+1))$     $subp$-$defn$
         $==$ if $i = i - (n+1)$ then 0 else $subp(i, i-(n+1)+1)+1$
    infer $subp(i, i-(n+1)) == n+1$     $==$-$subs(5.6, 5.5)$
6     $\forall n \in \mathbb{N} \cdot subp(i, i-n) == n$     $\forall$-$I(\mathbb{N}$-$ind(4,5))$
7     from $i \geq j$
7.1       $(i-j) \in \mathbb{N}$     $\mathbb{N}$, h7
    infer $subp(i,j) == i-j$     $\forall$-$E(6, 7.1)$, $\mathbb{Z}$
infer $i \geq j \Rightarrow subp(i,j) == i-j$     $\Rightarrow$-$I(7)$

Figure 1: Proof about *subp* using strong equality

to a reader who views the implication as guarding the evaluation: it is after all true that, for appropriate values, $subp(i,j) = i-j$ (with the weak equality).

Nor does the problem end here, intuitively clear properties such as:

$\forall i,j \in Z \cdot i > j \implies subp(i,j) > 0$
$\forall i,j \in Z \cdot i \geq j \implies subp(i,j) \in N$
$\forall i,j \in Z \cdot i \geq j \implies subp(i+1,j) \neq subp(i,j)$ etc.

would all need to be reformulated before proof could even be considered. Such reformulation could be handled by introducing strong versions of their operators. Alternatively it can be done systematically by reading any term $t$, for which a Boolean value is required, as though it were $t ==$ true. But this process of redefining 'innermost' terms must be handled with care. One might expect $subp(i,j) = i-j \lor subp(i,j) \neq i-j$ to be true but an interpretation using $(subp(i,j) \neq i-j) ==$ true would yield false for some values. (The expected result requires that $\neq$ is carefully interpreted such that the above expression becomes $subp(i,j) = i-j \lor \neg(subp(i,j) = i-j)$ before the other rewriting is handled.) Furthermore, the rewriting in the case of $subp(i,j) < i-j \lor subp(i,j) \geq i-j$ is even less clear. Another problem with applying the rewriting to innermost terms is that hd [ ] = 5 $\implies$ hd [ ] < 3 becomes true!

The preservation of the classical logical operators has been purchased at the cost of more complex relational operators. As an alternative, it is possible to view the 're-reading' as being applied only where a truth value is required by another context (e.g. the test of a conditional). Unfortunately this re-writing of 'outermost' terms leaves the problem of a collection of Boolean operators which handle undefined.

It was pointed out to one of the current authors (CBJ) by Oliver Schoett that strong equality is not the only operator which could be used to preserve the classical two-valued calculus. *Existential equality* yields false if either operand is undefined; it is again not strict; nor is it monotonic; it differs from strong equality only in that it is false if both operands are undefined. Existential equality is used in PROSPECTRA [Kri88] and would yield proofs similar to those in Figure 1 and difficulties similar to those for ==. In fact, it is possible to define:

$e_1 == e_2$ iff $e_1 =_3 e_2 \lor (\neg(e_1 =_3 e_1) \land \neg(e_2 =_3 e_2))$

Also:

$(e_1 = e_2) ==$ true iff $e_1 =_3 e_2$

As can be seen, existential equality has the additional disconcerting property that it is not reflexive and $\neg(e_1 =_3 e_2)$ is true if either operand is undefined.

## 5 Conditional operators

The difficulties seen in the preceding sections suggest that it is at least worth considering a change to the meaning of the logical operators so that they handle undefined values. Section 2 has already mentioned the idea of using the conditional expression form as an operator. This non-strict (in its first argument) form can be used to reformulate properties such as:

$\forall i,j \in Z \cdot$ if $i \geq j$ then $subp(i,j) = i-j$ else true
$\forall t \in N^* \cdot$ if $t \neq [\,]$ then $t = append(\text{hd}\, t, \text{tl}\, t)$ else true

But this clearly presents a significant step away from normal predicate calculus proofs. Already in the 1960's, McCarthy suggested [McC67] staying close to predicate calculus by redefining all of the classical logical operators in terms of conditional expressions. Conditional forms of or/and/implies can be defined:

$E_1$ cor $E_2$     iff     if $E_1$ then true else $E_2$
$E_1$ cand $E_2$     iff     if $E_1$ then $E_2$ else false
$E_1$ cimpl $E_2$     iff     if $E_1$ then $E_2$ else true

The obvious reformulations of the properties of sequences, maps and *subp* all become valid. These operators are monotone in the ordering shown in Figure 2 which has the pleasing computational intuition that a result can be computed if the first operand provides enough information even if the computation of the second operand is incomplete; furthermore, the latter's completion cannot result in a contradiction to the computed value. Here, however (cf. next section), a left-to-right evaluation order is implied (this is sometimes referred to as a 'sequential interpretation').

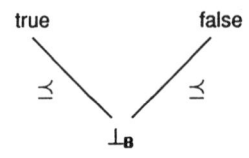

Figure 2: Ordering for Truth Values

With so many many pleasing attributes, why should one seek further? For any non-classical version of a logic, one has to ask which properties no longer hold. Here, the conditional forms of and/or do not enjoy one of the most intuitive properties expected of such operators: their operands can not be arbitrarily commuted. (Similarly, the contrapositive of an implication can fail.) In order to facilitate proofs, one of the current authors used both classical ($\wedge$) and conditional (&) forms in [Jon72]; a similar distinction is used (and vs. cand) in [Dij76]. (The programming language 'Standard ML' [HMT88] uses andalso.) None of these references, unfortunately, provides a full proof theory for the combined logic. What gives rise to greater concern is that there are some surprises in the identities. Whilst forms of de Morgan's law hold:

$\neg(E_1$ or $(E_2$ cand $E_3))$     iff     $\neg E_1$ and $(\neg E_2$ cor$\neg E_3)$

and cand left distributes over cor:

$E_1$ cand $(E_2$ cor $E_3)$     iff     $E_1$ cand $E_2$ cor $E_1$ cand $E_3$

more care is needed to see that the following property – used without comment in a program development in [Gri81] – holds:

$E_1$ and $(\neg E_1$ cor $E_2)$     iff     $E_1$ cand $E_2$

(Also, for example, cand does *not* right distribute over cor.) There is also a difficult decision to be made about the interpretation of quantifiers since there is no linear text to suggest the order of evaluation (this issue is scrutinized in [Bli88]).

On balance, this first attempt to adopt a logic which handles partial functions appears no easier to use than approaches which attempt to preserve classical logic. (This comment is made in the light of the experience in writing [Jon72].)

# 6  LPF

It is not difficult to discover other logics which handle partial functions but it should be clear from the preceding section that their acceptance will be decided on their algebraic properties. The particular 'Logic of Partial Functions' (hereafter, 'LPF') presented in [BCJ84] has a distinguished parentage – much of which was made known to the current authors after the publication of [BCJ84]. Kleene [Kle52] attributes a collection of symmetrical 'three-valued' operators to Łukasiewicz; Blikle [Bli88] traces this reference back to [Łuk20]; this logic was further studied in [Kol76] which was brought to the current authors' attention by Peter Aczel. But [Dri88] attributes such logics to MacColl and [Urq86] finds foreshadowings in the work of Boole, Pierce and Vasiliev but concedes that the work was focussed by Post and Łukasiewicz. One of the current authors developed the logic of [BCJ84] in his doctoral thesis ([Che86] formalises LPF as a first-order predicate calculus with equality in a sequent calculus with a set-theoretical semantics, upon which completeness and consistency are established. A cut-elimination theorem is proven.) The logic is used in VDM in [Jon86, Jon90, JS90]. This work has evoked some further interest [Gib88, Ten87, Avr88, Jer88, Hoo87] (Hoogewijs had studied related logics in [Hoo77]).

The basic idea is very simple: the propositional operators are given the strongest monotone (with respect to the ordering in Figure 2) extension of their two-valued interpretations. This removes the left-to-right order required for the conditional expression interpretation of the operators. Having abandoned order, it is natural to also give a 'generous' interpretation to the quantifiers (in terms of [Bli88], Kleene's quantifiers rather than McCarthy's); bound variables in quantified expressions range only over the 'proper' values of their bound set. The resulting logic has no positive surprises for the user since it employs only the normal collection of operators and all truths of LPF hold in classical first-order predicate calculus. Thus, for example, and/or are commutative and distribute over each other in the normal way. The negative surprises – which derive from the fact that the so-called 'law of the excluded middle' (*Tertium non datur*) does not hold – are discussed below.

Having elevated the *subp* result to the level of a challenge problem, the next step is to show its proof using LPF. This is done in Figure 3.[2] As in Section 3, the definition of *subp* is most conveniently handled by inference rules:

$$\text{subp-b } \frac{i \in \mathbb{Z}}{subp(i, i) = 0}$$

$$\text{subp-i } \frac{i_1, i_2 \in \mathbb{Z};\ i_1 \neq i_2;\ subp(i_1, i_2 + 1) = i_3}{subp(i_1, i_2) = i_3 + 1}$$

A simpler set of inference rules can be used for total functions. For partial functions, the rôle of the weak equality in the hypothesis of *subp-i* is important but one could envisage mechanical generation of such rules from recursive definitions. Notice that this proof only uses the weak equality required by the original statement of the property to be proved. Its only blemish appears to be the need (in line 7) to prove that the hypothesis is defined before the implication introduction rule can be used. Because the deduction theorem does not hold in general in LPF, the inference rule is:

$$\Rightarrow \text{-}I \frac{E_1 \vdash E_2;\ \delta(E_1)}{E_1 \Rightarrow E_2}$$

---

[2] Slightly different approaches – each using LPF – are given in [BCJ84] and [Jon86]. One other approach would be to proceed as in Figure 1 and then to use the inference rule which relates strong and weak equality.

Where $\delta(E)$ asserts that $E$ is defined (e.g. $E \vee \neg E$). The expressive completeness of the (re-interpreted) logical connectives for all possible monotone truth-valued functions (i.e. for $n \geq 0$, functions of type $\{\text{true}, \text{false}, \bot\}^n \to \{\text{true}, \text{false}, \bot\}$) is proved in [Kol76]. Thus this part of the logic is both formally and pragmatically expressive. See also [Kol81]. There are proofs – particularly of a meta-nature – which require non-monotonic truth-valued functions. Obviously, some non-monotonic connective has to be added in order to facilitate such expressions: LPF has a definedness connective ($\Delta$), which is first formalised in [Hoo77] and is sufficiently expressive (see also [Che86]; Blamey [Bla80] has pointed out the rôle of conditional expressions in giving a point, in a hierarchy of expressiveness, between strict and monotone operators). The interpretation of sequents in LPF is important: *models* are interpreted as binding the individual free variables in both the hypothesis and conclusion. A sequent is valid if its conclusion is true whenever its hypothesis is. Notice that, when the hypothesis is undefined, there is no constraint on the conclusion: it can be true, false or undefined (see Section 3.2 of [Avr87] for a discussion of 'consequence relations'). This differs from an implication and is the reason why the implication introduction rule has an additional hypothesis.

from $i, j \in \mathbb{Z}$
1    $i - 0 = i$                                                                                            h, $\mathbb{Z}$
2    $subp(i, i) = 0$                                                                                       h, $subp$-b
3    $subp(i, i - 0) = 0$                                                                                   = -$subs$(1,2)
4    from $n \in \mathbb{N}$; $subp(i, i - n) = n$
4.1      $i - (n + 1) \in \mathbb{Z}$                                                                       h, h4, $\mathbb{Z}$
4.2      $i \neq i - (n + 1)$                                                                               h, h4, $\mathbb{Z}$
     infer $subp(i, i - (n + 1)) = n + 1$                                                                   h, 4.1, 4.2, h4, $subp$-i
5    $\forall n \in \mathbb{N} \cdot subp(i, i - n) = n$                                                    $\forall$-$I$(N-ind(3, 4))
6    from $i \geq j$
6.1      $(i - j) \in \mathbb{N}$                                                                           N, h6
     infer $subp(i, j) = i - j$                                                                             $\forall$-$E$(5, 6.1), $\mathbb{Z}$
7    $\delta(i \geq j)$                                                                                     h, $\mathbb{Z}$
infer $i \geq j \Rightarrow subp(i, j) = i - j$                                                             $\Rightarrow$ -$I$(6,7)

Figure 3: Proof about *subp* in LPF

Does this difference bring any surprises to the user of LPF? Consider:

*evenp* : $\mathbb{Z} \to \mathbb{B}$

*evenp*($i$) $\triangleq$ if $i = 0$ then true else *evenp*($i - 2$)

With the obvious modulus operator:

*evenp*($i$) $\vdash$ $i$ mod $2 = 0$

is a valid sequent. If the inference rules are generated for *evenp* in the same way as for *subp* above, they are:

*evenp*-b $\dfrac{}{evenp(0)}$

$$\text{evenp-i} \frac{i \neq 0; \ evenp(i-2)}{evenp(i)}$$

But these are *not* sufficient to prove the sequent. They are, in fact, only requiring a fixed point of the definition rather than the least fixed point.[3] Thus $evenp(i) \Leftrightarrow$ true satisfies *evenp-b* and *evenp-i*. What is needed to prove a sequent whose hypothesis is $evenp(i)$ is an upper bound and this can be provided by an induction rule. It is not however obvious how such rules can be mechanically generated from the recursive function definitions. Such leastness inference rules are needed in other cases: for example, to show:

$subp(i,j) \neq i-j \vdash i<j$

But notice that the implication $\forall i,j \in Z \cdot subp(i,j) \neq i-j \implies i<j$ is the contrapositive of the 'challenge' result and follows therefrom without difficulty.

# 7  Owe's 'weak logic'

Owe's work [Owe85] is aimed at providing a logic for reasoning in program development: partial functions are identified with non-terminating subprograms or ones which terminate exceptionally. Motivated by the concept of data abstraction and functional abstraction, three requirements are proposed for such a logic:

1. reasoning about defined formulae should follow the rules of traditional first-order logic;

2. theorems proved at an abstract level should (at least mostly) be valid on a less abstract level; and

3. on an abstract level, specification or identification of so-called implementation dependent errors can be avoided.

With these requirements, 'weak logic' is formalised. There are strictly total connectives such as *and, or, implies* and *not*, as well as non-strict and non-commutative ones like *and then, or else, implies then* and *if-construct*. With such connectives, more traditional (non-strict) connectives such as $\wedge$, $\vee$ and $\implies$ can be defined.

The propositional formulae and universal quantifications are interpreted in a 'weak semantics'; informally, this means:

A formula is valid if it is true whenever it is defined.

As a result, if a formula $q$ is undefined, then the formula false *and* $q$ is undefined and therefore trivially valid, whereas false $\wedge q$ is defined and evaluated to false.

In contrast, the existential quantifier is interpreted in a 'strong semantics':

$\exists x \in T \cdot q$ is valid if there is a value of $x$ in $T$ such that $q$ is defined and true.

($\forall$ is non-monotonic!) To cope with partial functions a *domain predicate*, denoted as *in f*, is associated with each function $f$ and is interpreted as follows:

---

[3]Notice that, just as with the distinction between weak and strong equality, it is not correct to provide an axiom $evenp(i) \Leftrightarrow i = 0 \vee evenp(i-2)$.

$x$ *in* $f$ is true if $f(x)$ is defined, and false otherwise.

The definedness of an expression $e$ can also be specified by a formula, denoted as $DEF[e]$, which is evaluated to true when $e$ is defined and false otherwise. Proof rules are developed in conjunction with the semantics, and completeness and consistency established.

Application of Weak Logic to the *subp* example would seem, at first sight, to yield a short proof, as no special attention need be given to the undefinedness, which in Weak Logic can be treated as true. This is akin to a so-called 'partial correctness' proof in program proofs. However, a separate proof is needed to establish the definedness, as proving termination for 'total correctness' of programs.

## 8 LCF

This section looks at the approach adopted by LCF – a very well-known and influential system. Based on Scott's domain theory, LCF [MMN75, GMW79] adopts layered-logic approach: separating the computable objects from the logic for reasoning about the objects.

The deductive calculus used in LCF is called *PP-lambda* [MMN75, GMW79]. Terms of *PP-lambda* are those of a typed $\lambda$-calculus with a fixed-point combinator, and formulae are:

**atomic formulae** an equality (==) and a partial ordering (<<) between terms,

**conjunction** normal (classical) $\wedge$,

**implication** normal (classical) $\Rightarrow$ , and

**Universal Quantification** $\forall$ ranges over a *domain* (including a bottom element $UU$, which is a term).

All formulae are supposed to be two-valued, and in particular, $UU == UU$ is evaluated to $TRUTH$.

As to the inference rules of *PP-lambda*, logical rules are basically those classical rules concerning $\wedge$, $\Rightarrow$ and $\forall$. Non-logical rules deal with the equality, the partial ordering, $\lambda$-conversions and the fix-point combinator.

Because of the range of quantifiers, we would need, in the *subp* example, to establish in development of the theory of natural numbers an induction principle as follows:

$$N\text{-}ind \quad \frac{p(UU); \quad p(0); \quad x > 0, p(x-1) \vdash p(x)}{\forall x \in N \cdot p(x)}$$

assuming $N$ is the domain of natural numbers with $UU$ embedded. Once this principle has been established (possibly using fixed-point induction), a proof of *subp* in *PP-lambda* is similar to the one in LPF. However, the equality must be strong, and an extra basis step for the induction performed (i.e. we need to prove $subp(i, i - UU) == UU$). This can be established with the non-logical rules. Nevertheless, all proofs appear to involve lines of deduction dealing with the undefined value $UU$.

Blikle in [Bli88] also considers a 'layered approach', where the 'superpredicates' are used as a bridge between three-valued predicates and the classical logic. Four conceptual levels are distinguished:

1. the level of a software system;
2. the level of a logic of a software system;
3. the level of a definitional metalanguage; and
4. the level of a logic of a definitional metalanguage.

## 9 Plotkin's 'PFL'

Similar to *PP-lambda* of LCF, Plotkin's PFL (Partial Function Logic) [Plo84] is strongly influenced by Scott's domain theory and is intended to be used to reason about computable functions. But unlike *PP-lambda*, the logic itself is a free logic similar to Scott's [Sco79] (see also [Fef82] and the discussion of 'LPT' in [Bee86]; 'COLD-K' [Jon88, FJO+87, FJ87] also uses a logic based on Scott's ideas). The propositional inference rules are normal. However, because in PFL free variables may not denote but bound variables always do, we have the following rules for the quantifiers:

$$\forall\text{-}I \frac{x\downarrow \;\vdash\; p(x)}{\forall x \cdot p(x)}$$

$$\forall\text{-}E \frac{\forall x \cdot p(x); \; e\downarrow}{p(e)}$$

$$\exists\text{-}I \frac{p(e); \; e\downarrow}{\exists x \cdot p(x)}$$

$$\exists\text{-}E \frac{p(x), x\downarrow \;\vdash\; q; \; \exists x \cdot p(x)}{q}$$

where $x\downarrow$ means '$x$ is defined', and similarly for $e\downarrow$.

Due to this treatment of free and bound variables, many references to the undefined can be avoided. For example, in Plotkin's axiomatization of natural numbers, with Scott Induction, the normal induction schema for natural numbers can be derived.

This implies that, for the *subp* example, a proof in PFL would be similar to the one in LPF. The similarity may be just superficial. Underneath, PFL is based on a fairly sophisticated semantics handling ordering, existence, a partial typed $\lambda$-calculus, and many other notions.

## 10 Related papers

In [KTB88] the authors present a three-valued logic, using McCarthy's *CAND* and *COR* and Kleene's quantifiers. Different notions of validity (strong *always true* and weak *never false*) are distinguished; a classification of consequence relations in model-theoretic terms according to different treatments of models and theorems is given (e.g. a logic can have strong models and weak theorems). Several logics are compared with these notions. The proof theory is developed with Beth's semantic tableaux and a transformation to sequent and natural deduction systems. (Similar ground is covered in [GFLC89].)

As is mentioned in Section 3, various approaches to handling partial functions are included in the literature on 'Z'. The main semantics in [Spi88] actually results in an existential equality interpretation (cf. Section 4 above) but Section 4.3 of [Spi88] also contains an excellent discussion of alternatives and a conclusion which indicates why LPF might be thought of as more suitable for VDM whilst 'Z' requires a different approach. He also draws attention to [Abr84a] and its use of 'intentionally incomplete definition of the semantics of predicates'. It would be interesting to see a comparison of this approach with the 'supervaluations' which [Urq86] traces to [vF66] ([Ben86] lists some problems of this approach).

A thorough discussion of logics which cope with undefined terms is contained in [Bla80] or the more accessible [Bla86]. In [Hoo87], a comparison of LPF and the logic PPC (partial-predicated calculus) of [Hoo77] is given; the usefulness of the non-monotone connective is demonstrated. A discussion – in a linguistic setting – of 'free logics' is given in [Ben86]; '$L_3$' is proposed in [Sch87] for the study of natural language semantics. The review of approaches in [Avr88] is far easier to relate to the problems of program proofs.

Broy [Bro87a] discusses equational specification of partial (and higher-order) algebras; the paper both formalises 'partial interpretations' and presents a proof theory. In [Bro88], Broy argues for the use of partial algebras because of the messiness of handling error values at the specification level.

A paper which discusses the problems of importing expressions from a programming language into the proof logic (as happens, for example, in Hoare-logic) is [Ten87]. The approach uses quantifiers which range over all (including $\bot$ values); so-called 'strict interpretations' (equivalent to what has been called above 'existential equality'); and a distinguished equality ('strong' in the terms used here) for axioms. Special rules are provided for the manipulation of assertions like ($p$ and $q$) = false. Jervis also considers partial terms in Hoare-like logics. In particular, [Jer88] investigates the use of LPF. Amongst other interesting results, his Theorem 2.3.8 shows the existence of normal forms for expressions of LPF and his distinctions on sequent notions versus implications are illuminating. Tucker and Zucker argue that weak (strict) logical operators are appropriate for their exhaustive treatment [TZ88] of error-state semantics. Further references (for which the authors are grateful to Professor Ito – whose, as yet unpublished, 'Logic of Execution' is also relevant) are [Sat87, HN88].

# 11 Conclusions

It would be dangerous for the reader to expect the proponents of one of the approaches to provide a completely balanced consumer survey of the alternatives discussed in Sections 2–9. It is then fairer to state at the outset that this concluding section describes the reasons that the current authors believe that their own approach (cf. Section 6) satisfies the main criteria for reasoning about partial functions in software development.

For any logic, it is clearly desirable to have both model and proof theories; the latter must be consistent with the former and should preferably also be complete (for LPF, these results are covered in [Che86]); other non-standard logics frequently lack a proof theory. The difficulty of using a non-classical logic can be minimised in two ways. The valid sequents of LPF are all consistent with the classical (two-valued) model. Furthermore, any classical tautology can be made true in LPF simply by introducing definedness assumptions for each of its propositions. Conservation in LPF of properties like symmetry of conjunctions and disjunctions and the appropriate distributivity laws could also be regarded as a benefit. Additionally, there would appear to be a strong computational intuition in requiring that the non-strict operators are the strongest monotone (with respect to the ordering in Figure 2) extension of the classical operators.

LPF preserves the (classical) link between $p \Rightarrow q$ and $\neg p \vee q$. Łukasiewicz's truth table for implication gives true for the $\bot \Rightarrow \bot$ case and Wang [Wan61] argues that $p \Rightarrow p$ must be an identity which, in a logic without the law of the excluded middle, is incompatible with the usual definition of implication. The decision in this area is independent of most of the others in LPF and could be changed with little impact.

One objection to LPF is that it is non-standard and unfamiliar ([Fef82] argues against 'three-valued and other restrictive logics due to their debilitating effects on ordinary reasoning'!). In fact, the key difference from classical logic is the lack of the law of the excluded middle. In itself, this is unlikely to cause problems: it is quite valid to introduce (from knowledge of the data structures) a property like: $i > j \vee i \leq j$ (where both $i$ and $j$ are integers); banning disjunctions like: $i/0 = 5 \vee i/0 \neq 5$ is unlikely to be limiting. The indirect consequences (e.g. the inability to reason by *reductio ad absurdum*) are more difficult to assess. It is likely that fully formal proofs will only come about with the availability of usable Theorem Proving Assistants (see, for example, [JL88]). In this case, the unfamiliarity of a particular set of proof rules is likely to matter much less than other factors in the design of the User Interface.

Those approaches which avoid the need for a non-classical logic either bring their own surprises into proofs or force the separation of validity proofs from new ones for definedness. It has been a major goal in the development of LPF itself, and ways of using it in program development, that the normal user (as distinct from someone proving meta-results) should not have to reason about undefined values nor use an additional notion of strong equality.

One last objection which is levelled at LPF (mentioned for example in [Bli88]) is that it is 'unimplementable'. This point, in spite of its pragmatic tone, appears to be of minor practical significance. The property of *subp* which has been a *leitmotiv* of this paper arises out of the desire to prove that the recursive definition of *subp* satisfies a specification given by pre- and post-conditions. Those logical operators which do occur within, say, post-conditions are normally eliminated during design. (Consider, for example, a specification of *sort* which requires *is-ordered*(*outl*) ∧ *is-permutation*(*inl, outl*): the conjunction is a specification trick and not intended to be part of the implementation; similar cases arise with the use of invariants). The rather small percentage of logical operators which might end up being translated into tests of conditionals in a programming language, should require only simple checking to avoid reliance on LPF's generous treatment of undefined (see if-*I* and while-*I* in [Jon90]).

# Acknowledgements

The authors are grateful to Peter Lindsay for his comments on a draft of this paper. Thanks also go to members of IFIP's WG2.2 and WG2.3 working groups (especially Broy and Loeckx) with whom the evolving LPF has been discussed. One of the authors (CBJ) would like to thank Professors Burstall and Ito for arranging the Sendai workshop (where this material was first presented) and the Japanese attendees for their incredible hospitality. Thanks are also due to SERC who funded both the trip to Japan and the related research (by a Research Grant and a Senior Fellowship) and to the Wolfson Foundation for their research grant. The other author (JHC) would like to thank his employer for support.

# References

[Abr84a]　J-R. Abrial. The mathematical construction of a program and its application to the construction of mathematics. *Science of Computer Programming*, 4:45–86, 1984.

[Abr84b]　J-R. Abrial. *Programming as a Mathematical Exercise*, pages 113–137. Prentice-Hall International, 1984.

[Abr87]　J.R. Abrial. Private communication. Letter dated May 31st., 1987.

[Avr87]　A. Avron. Simple consequence relations. Technical Report ECS-LFCS-87-30, LFCS, Department of Computer Science, University of Edinburgh, June 1987.

[Avr88]　A. Avron. Foundations and proof theory of 3-valued logics. Technical Report ECS-LFCS-88-48, LFCS, Department of Computer Science, University of Edinburgh, April 1988.

[BCJ84]　H. Barringer, J.H. Cheng, and C.B. Jones. A logic covering undefinedness in program proofs. *Acta Informatica*, 21:251–269, 1984.

[Bee86]　M.J. Beeson. Proving programs and programming proofs. In B. Marcus et al., editors, *Logic, Methodology and Philosophy of Science*, volume VII, pages 51–82. Elsevier, 1986.

[Ben86]　E. Bencivenga. Free logics. In D. Gabbay and F. Guenthuer, editors, *Handbook of Philosophical Logic, Volume III*, chapter 6. Reidel, 1986.

[BJ82]　Dines Bjørner and Cliff B. Jones. *Formal Specification and Software Development*. Prentice Hall International, 1982.

[Bla80]　S.R. Blamey. *Partial Valued Logic*. PhD thesis, Oxford University, 1980.

[Bla86]　S. Blamey. Partial logic. In D. Gabbay and F. Guenthuer, editors, *Handbook of Philosophical Logic, Volume III*, chapter 1. Reidel, 1986.

[Bli88]　A. Blikle. Three-valued predicates for software specification and validation. In R. Bloomfield, L. Marshall, and R. Jones, editors, *VDM—The Way Ahead*, pages 243–266. Springer-Verlag, 1988. Lecture Notes in Computer Science, Vol. 328.

[Bro87a]　M. Broy. Equational specification of partial higher order algebras. In M. Broy, editor, *Logic of Programming and Calculi of Discrete Design — NATO ASI Series F: Computer and Systems Sciences, Vol. 36*, pages 185–241. Springer-Verlag, 1987.

[Bro87b]　M. Broy. Predicative specifications for functional programs describing communicating networks. *Information Processing Letters*, 25:93–101, 1987.

[Bro88]　M. Broy. Views of queues. *Science of Computer Programming*, 11:65–86, 1988.

[Che86]　J.H. Cheng. *A Logic for Partial Functions*. PhD thesis, University of Manchester, 1986.

[Dij76]　E.W. Dijkstra. *A Discipline of Programming*. Prentice-Hall, 1976. In Series in Automatic Computation.

[Dri88]　D. Driankov. Towards a many-valued logic of quantified belief. *Technical Report 192*, Linköping Studies in Science and Technology, 1988.

[Fef82]　S. Feferman. Toward useful type-free theories, I. *Journal of Symbolic Logic*, pages 75–111, 1982.

[FJ87]　L.M.G. Feijs and H.B.M. Jonkers. Formal definition of the design language COLD-K. Technical report, Philips Research Labs, The Netherlands, April 1987. Preliminary Edition.

[FJO+87]　L.M.G. Feijs, H.B.M. Jonkers, J.H. Obbink, C.P.J. Koymans, G.R. R.de Lavaletter, and P.H. Rodenburg. A survey of the design language COLD. In *ESPRIT '86: Results and Achievements*, pages 631–644. Elsevier Science Publishers B.V., 1987.

[GFLC89]　A. Gavilanes-Franco and F. Lucio-Carrasco. A first order logic for partial functions. Technical report, Universidad Complutense, Madrid, 1989.

[Gib88]　P.F. Gibbins. VDM: Axiomatising its propositional logic. *BCS Computer Journal*, 31:510–516, 1988.

[GMW79]　M. Gordon, R. Milner, and C. Wadsworth. *Edinburgh LCF*, volume 78 of *Lecture Notes in Computer Science*. Springer-Verlag, 1979.

[Gri81]　D. Gries. *The Science of Computer Programming*. Springer-Verlag, 1981.

[Hay87]　I. Hayes, editor. *Specification Case Studies*. Prentice-Hall International, 1987.

[HMT88]　R. Harper, R. Milner, and M. Tofte. The definition of standard ML: Version 2. Technical report, Laboratory for Foundations of Computer Science, Edinburgh University, 1988.

[HN88]　S. Hayashi and H. Nakano. *PX: A Computational Logic*. MIT Press, 1988.

[Hoo77]　A. Hoogewijs. A calculus of partially defined predicates. Technical report, Rijksuniversiteit, Gent, 1977.

[Hoo87]　A. Hoogewijs. Partial-predicate logic in computer science. *Acta Informatica*, 24:381–393, 1987.

[Jer88]　C.A. Jervis. *A Theory of Program Correctness with Three Valued Logic*. PhD thesis, Leeds University, 1988.

[JL88]　C.B. Jones and P.A. Lindsay. A support system for formal reasoning: Requirements and status. In R. Bloomfield, L. Marshall, and R. Jones, editors, *VDM'88: VDM— The Way Ahead*, pages 139–152. Springer-Verlag, 1988. Lecture Notes in Computer Science, Vol. 328.

[Jon72]　C.B. Jones. Formal development of correct algorithms: an example based on Earley's recogniser. In *SIGPLAN Notices, Volume 7 Number 1*, pages 150–169. ACM, January 1972.

[Jon80]　C.B. Jones. *Software Development: A Rigorous Approach*. Prentice Hall International, 1980.

[Jon86] C.B. Jones. *Systematic Software Development Using VDM*. Prentice Hall International, 1986.

[Jon88] H.B.M. Jonkers. An introduction to COLD-K. Technical Report METEOR/t8/PRLE/8, Philips Research Labs, Eindhoven, July 1988.

[Jon90] C. B. Jones. *Systematic Software Development using VDM*. Prentice Hall International, second edition, 1990.

[JS90] C.B. Jones and R.C.F. Shaw, editors. *Case Studies in Systematic Software Development*. Prentice Hall International, 1990.

[Kle52] S.C. Kleene. *Introduction to Metamathematics*. Van Nostrad, 1952.

[Kol76] G. Koletsos. Sequent calculus and partial logic. Master's thesis, Manchester University, 1976.

[Kol81] G. Koletsos. Notational and logical completeness in three-valued logic. *Bull. of the Greek Mathematical Society*, 22:121–141, 1981.

[Kri88] B. Krieg-Brückner. The PROSPECTRA methodology of program development. In Zalewski, editor, *IFIP/IFAC Working Conference on Hardware and Software for Real-Time Process Control*, page ?? ??, 1988.

[KTB88] B. Konikowska, A. Tarlecki, and A. Blikle. A three-valued logic for software specification and validation. In R. Bloomfield, L. Marshall, and R. Jones, editors, *VDM— The Way Ahead*, pages 218–242. Springer-Verlag, 1988. Lecture Notes in Computer Science, Vol. 328.

[Lin87] P.A. Lindsay. A formal system with inclusion polymorphism. Technical Report IPSE Document 060/pal014/2.3, Manchester University, December 1987.

[Łuk20] J. Łukasiewicz. O logice trójwartościowej. *Ruch Filozoficzny*, pages 169–171, 1920. Translated as (On three-valued logic) in Polish Logic 1920–39, S. McCall (ed.), Oxford U.P., 1967.

[Man74] Z. Manna. *Mathematical Theory of Computation*. McGraw-Hill, 1974.

[McC67] J. McCarthy. A basis for a mathematical theory for computation. In P. Braffort and D. Hirschberg, editors, *Computer Programming and Formal Systems*, pages 33–70. North-Holland Publishing Company, 1967.

[MMN75] R. Milner, L. Morris, and M. Newey. A logic for computable functions with reflexive and polymorphic types. In *Proceedings of Conference on Proving and Improving Programs*, 1975.

[Owe85] O. Owe. An approach to program reasoning based on a first order logic for partial functions. Technical Report 89, Institute of Informatics, University of Oslo, February 1985.

[Plo84] G.D. Plotkin. Types and partial functions. Seminar at University of Manchester., 1984.

[Sat87]  M. Sato. QJ. In *Fourth International Conference on Logic Programming*, 1987.

[Sch86]  D.A. Schmidt. *Denotational Semantics: a Methodology for Language Development.* Allyn & Bacon, 1986.

[Sch87]  P.H. Schmidt. Computational aspects of three-valued logic. Technical Report LILOG 26, IBM Deutschland, 1987.

[Sco79]  D. S. Scott. Identity and existence in intuitionistic logic. In *Lecture Notes in Mathematics (Volume 735)*. Springer-Verlag, 1979.

[Spi88]  J.M. Spivey. *Understanding Z—A Specification Language and its Formal Semantics.* Cambridge Tracts in Computer Science 3. Cambridge University Press, 1988.

[Sto77]  J.E. Stoy. *Denotational Semantics: The Scott-Strachey Approach to Programming Language Theory.* MIT Press, 1977.

[Ten87]  R.D. Tennent. A note on undefined expression values in programming logic. *Information Processing Letters*, 24(5), March 1987.

[TZ88]  J.V. Tucker and J.I. Zucker. *Program Correctness over Abstract Data Types, with Error-State Semantics.* CWI Monograph. North-Holland, 1988.

[Urq86]  A. Urquhart. Many-valued logic. In D. Gabbay and F. Guenthuer, editors, *Handbook of Philosophical Logic, Volume III*, chapter 2. Reidel, 1986.

[vF66]  B.C. van Fraasen. Singular terms, truth-value gaps and free logic. *J. Philosophy*, 63:481–495, 1966.

[Wan61]  H. Wang. The calculus of partial predicates and its extension to set theory. *Math. Logic*, 7:283–288, 1961.

# Morris

One of the benefits of having refinement techniques for programs is the backwards pressure they exert on the means by which specifications themselves are made. From early experience with large-scale specifications it was obvious that they must be structured: only by making large specifications from combinations of small ones can the whole be understood.

Knowing that structuring of specifications is important is not enough, unfortunately, to deliver the precise details of the way in which the structuring tools should work. Yes, conjunction is required — specifying that the product must do this *and* this — but what precisely is conjoined, and how? Deciding that arbitrarily, and then attaching a refinement method, runs the risk of encountering left-handed nuts but right-handed bolts at the very end.

The nuts and bolts of program development generally are the rules with which small intricate programs are developed. Naturally most applications contain proportionally very little intricacy; but that little part, however small, is just as important for correctness as are the sweeping decisions made about large-scale structure.

Knowing more now about the refinement of small programs it is possible to go back to the question of large-scale structure and make adjustments where necessary to ensure the uniformity of the refinement method overall. In this paper the author blends knowledge of small-scale concerns with the well-known requirements for large-scale structuring, showing thus how specifications might be made in the first place.

# Designing and Refining Specifications with Modules

Joseph M. Morris and Shahad N. Ahmed [*]

## Abstract

Specifications have a dual existence: on the one hand they constitute a contract between the customer and the vendor, and on the other hand they are the instructions from the vendor to his programmers. From the former viewpoint, we would like to have methods for incrementally constructing specifications, and from the second viewpoint we would like to have mechanisms for systematically "calculating" an implementation of a specification. Here we seek to develop a methodology that can give formal support to both these activities within a model-oriented framework. Indeed, these two activities, which we may call designing and refining, have much in common, and to some extent we may also "calculate" parts of a specification. The basic theoretical notion underlying our approach is that programs and specifications are more or less the same thing: a programming language is the implementable subset of a specification language. The basic practical tool we outline is a notion of modules that supports the incremental construction of specifications.

## 1. Introduction

Specifications play a dual role. On the one hand they constitute the contract between the customer and the owner of the software company, and on the other they are the instructions from the owner to his programmers. Work on specifications has mainly been from the viewpoint of specifications as contracts, and their role in program construction is not nearly as well developed or understood. Typically a programmer will read a specification to the extent that he needs to know what is being asked of him, and then gets on with the job of programming. Perhaps he will refer back to the specification in moments of doubt, but it will not otherwise infringe on his time. But we well know by now that the specification can usefully play a much more active role in programming: it is possible, at least in principle, to derive a program from its specification by mathematical transformation. Our purpose here is to outline a specification methodology that recognises and supports the dual role of specifications in designing systems and programming by transformation.

Our basic tenet is that programs and specifications are more or less the same thing: a programming language is the implementable subset of a specification language, or viewed the other way around, a specification language is a programming language with

---
[*] Dept of Computing Science, University of Glasgow, Glasgow G12 8QQ, Scotland, U.K.

added fancy constructs that admit ease of expression although they may be expensive or even impossible to implement. The programming task is to make a specification in this rich language and then step by step eliminate the fancy constructs with correctness-preserving transformations — so called "refinements" — until we arrive at a program. Programs *are* specifications, although somewhat special in that we know how to execute them. Although occasionally the best specification of a program is the program itself, usually the specification will be quite un-program-like; then we have to develop the program by constructing a sequence of ever more algorithmic specifications until we arrive at one from which all the non-algorithmic constructs have been eliminated. This is pretty much what we do informally when we program by stepwise refinement.

Specifications are as complex in their structure as programs. We have to construct them very methodically in small pieces which we then combine into bigger specifications, and so on hierarchically. The specification language Z [7] has pioneered this style. Coincidentally, programming by stepwise refinement also proceeds by repeatedly translating small pieces of specification into code. As far as is practicable we would like the constituent pieces of the specification also to be the pieces we make the subject of refinements, or put more simply, we would like the program to have much the same structure as the specification. The advantages are these. Firstly, we avoid the effort of reshaping the specification prior to deriving its implementation. Secondly, it places a tight upper bound on the amount of recoding we have to do when a part of the specification is changed. Thirdly, we can more easily reuse implementations when a piece of a specification is used more than once. We regard this latter point as important in that it encourages us to make libraries of useful specifications, together with their implementations, that can be used in more than one application. Although it seems desirable that the program should retain much of the structure of the specification, in fact it is not always easy to achieve nor does it always yield an acceptably efficient implementation. Nonetheless we would like to encourage it.

The presentation will be tutorial in style. We assume some idea on the readers part of the purpose of specifications, but not any great intimacy with them. We will proceed mainly by example, without attempting to describe the mathematical underpinnings of the method. Of course it is very important that the specification language be given a formal semantics so that we can formally derive the laws of refinement, but that is a separate issue and is discussed elsewhere [1-6]. We use various typefaces in specifications and explanatory text, just to help the eye; they have no significance otherwise.

## 2. A simple example

Figure 1 is a specification in a Pascal-like syntax of a simple library catalogue, with explanations following. The actual type definitions of BOOK etc. aren't significant and are left open. Each copy of a book is associated with a unique identifier, so for BOOK we would probably choose the naturals, TITLE will probably be a sequence of characters, and so on. Because we are writing a specification, which doesn't have to be implemented, we can avail of any mathematical types we find useful. The values in type "set of BOOK" are all subsets of the values in BOOK. Variable *title* has type BOOK $\nrightarrow$ TITLE — in words, *title* is a partial function from BOOK to TITLE; its values are sets of pairs, the left hand element of which is from BOOK and the right hand element of which is from TITLE, and such that the left hand elements are unique. So *title* just associates each book identifier with a title. Similarly, *authors* associates each book identifier with its list of authors. The set of left hand elements in a partial function f is called the domain of f and is denoted by dom.f; the set of right hand elements is called the range of f and is denoted by rng.f. We will often want to constrain the values a variable may take on, over and above that imposed by its type; we library_catalogue module =

    **type**    BOOK = ... ;
              TITLE = ... ;
              AUTHOR = ... ;
              SUBJECT = ... ;

    **var**    books: set of BOOK;
              title: BOOK ↠ TITLE | books ⊆ dom.title;
              authors: BOOK ↠ sequence of AUTHOR | books ⊆ dom.authors;
              subject: BOOK ↠ SUBJECT | books ⊆ dom.subject;

    **procedure** addbook(b: BOOK; t: TITLE; as: sequence of AUTHOR; s: SUBJECT);
    **begin**
              **if** b∉ books **then**
              **begin**
                    books:= books ∪ {b}; title:= title ⊕ {(b,t)}
                    authors:= authors ⊕ {(b,as)}; subject:= subject ⊕ {(b,s)}
              **end**
              **else**   message('Non-unique book identifier')
    **end;**

    **procedure** removebook(b: BOOK);
    **begin**
              **if** b∈ books **then** books:= books - {b}
              **else** message('Unknown book')
    **end;**

    **function** booksby(a: AUTHOR): set of BOOK;
    **begin**
              booksby:= {b: b∈ books ∧ a∈ authors.b}
    **end;**

    **function** bookson(s: SUBJECT): set of BOOK;
    **begin**
              bookson:= {b: b∈ books ∧ s=subject.b}
    **end;**

**end**

Fig. 1: Library catalogue in Pascal-like syntax

include such constraints in clauses beginning with a vertical line. For example, we intend that *title* should include the title of every book currently in the library and we write this as books ⊆ dom.title. We might have expected to see books = dom.title as the constraint in this case; that is not excluded, of course, and will most likely be observed by the implementation, but it's not essential to the specification and so we choose the weaker constraint — specifications should be as unconstrained as possible while remaining consistent with the requirements. The conjunction of all such constraints is called the "invariant" of the module. We are free to use all the standard operations that come with types, such as set membership, set union, and set difference in the case of set types. The operation ⊕ on ordered pairs is so-called "function overriding": the result of title ⊕ {(b,t)}, for example, is a copy of *title* from which any pairs whose first element is b (there will be at most one in the present case) have been removed and to which {(b,t)} has been added. So title:= title ⊕ {(b,t)} just records the title $t$ of a new book $b$ being added to the library. The set membership symbol ∈ is extended to sequences in the obvious way. We shall not be too specific about handling "error situations"; for simplicity we will assume some message is to be conveyed, and assume the availability of routine "message" which achieves this. Function application is denoted by an infix dot; it has the highest operator precedence.

The library catalogue meets all the requirements of a good specification: it is precise, unambiguous, and says no more than is necessary to convey what is to be achieved. We are also told that specifications should not be operational, and on this ground the reader is perhaps feeling uneasy. Doesn't the specification use an assignment statement, for example, and isn't that an operational construct? Not at all! The assignment statement is just a convenient mathematical notation, simple and succinct. It so happens that it can also be given an operational interpretation, but that's just by the way as far as we're concerned. Note also that when removing a book from the catalogue we didn't specify any change to variables *title*, *authors*, or *subject*, and so these variables retain old information. We could have updated them if we cared to, but it is just extra work to no purpose: there is no notion of "space efficiency" of specifications, just as there is no notion of "time efficiency". The important thing is that a specification should not exclude efficient implementations, and it can be shown that the specification we have made does not. We will discuss implementations later. The specification of Figure 1 is complex in that it is composed of smaller specifications such as **addbook** and **removebook**, and these in turn are composed of still smaller specifications such as books:= books ∪ {b}.

Actually we do not use a Pascal-like syntax when we make specifications: the specification we would actually write is shown in Figure 2. To begin with, Pascal-like syntax is too verbose; there is a lot of writing in making specifications, and we prefer to be succinct. We make parameters of the types that need not be fixed in the module. The end of a parameter list in the definition of an operation is indicated by a period (which is quite different from the period of function application), and the end of a definition is indicated by a fat dot. The use of semicolons in Figure 1 is over-specific. It is not attractive to regard the removal of a book as consisting of four small actions carried out in some arbitrary order, for we really don't care what order they're carried out in, and they could even be carried out simultaneously. In general we avoid composing statements with semicolons, and instead use the parallel combinator ∥ : we shall describe this more precisely below, but for the moment it suffices to think of it as indicating that all the operations described by its arguments should take place in parallel. The commonly occurring form x:= x **op** y, where **op** is some operation, is abbreviated to x:**op** y; to avoid ambiguity this abbreviation is not used when **op** is =.

An if-statement is composed of a collection of so-called "guarded commands" combined together using the choice operator []. A guarded command has the form $P \rightarrow s$ where $P$ stands for an assertion (i.e. a boolean-valued expression) called the guard, and s stands for a specification. If $P$ is true then $P \rightarrow s$ behaves like s, and otherwise it doesn't behave at all: in that case we hope there is some other guarded command in the collection whose guard is true. A set of guarded commands specifies a choice among any one of those operations in the set that are preceded by a guard that evaluates to true. If more than one guard is true then it is not defined which corresponding operation is chosen: it is up to the implementor. For example,

$x \geq y \rightarrow y:=y+1$ [] $x \leq y \rightarrow x:=x+1$

leaves it open as to whether y:=y+1 or x:=x+1 is chosen when x=y. Specifications library_catalogue module [BOOK, TITLE, AUTHOR, SUBJECT] =

    *books*: set of BOOK;
    *title*: BOOK $\twoheadrightarrow$ TITLE | books $\subseteq$ dom.title;
    *authors*: BOOK $\twoheadrightarrow$ sequence of AUTHOR | books $\subseteq$ dom.authors;
    *subject*: BOOK $\twoheadrightarrow$ SUBJECT | books $\subseteq$ dom.subject;

    **addbook** =
        *b*: BOOK, *t*: TITLE, *as*: sequence of AUTHOR, *s*: SUBJECT.
          $b \notin$ books $\rightarrow$
              books:$\cup$ {b} || title:$\oplus$ {(b,t)} || authors:$\oplus$ {(b,as)} ||
              subject:$\oplus$ {(b,s)}
        [] $b \in$ books $\rightarrow$
              message('Non-unique book identifier') •

    **removebook** =
        *b*: BOOK.
          $b \in$ books $\rightarrow$ books:- {b}
        [] $b \notin$ books $\rightarrow$ message('Unknown book') •

    **booksby** =
        *a*: AUTHOR. {b: b$\in$ books $\wedge$ a$\in$ authors.b} •

    **bookson** =
        *s*: SUBJECT. {b: b$\in$ books $\wedge$ s=subject.b} •

**end**

Fig. 2: Library catalogue version 2

are in this way "nondeterministic" in that they describe not one unique computation but a collection of possible implementations all of which are acceptable to the customer. Nondeterminacy is an attractive property of specifications because it allows us to be non-committal when we really don't care; the implementor can then make a choice based on convenience or efficiency. The collection of implementations admitted by a specification could even be empty for some states: this would happen, for example, if there is some state for which none of the guards in a guarded command set is true. Such a specification is semantically okay, but unimplementable for those states, and we say it is "partial" or "miraculous" — "miraculous" because the specifier appears to be expecting miracles from the implementor. For example,

$$x>y \rightarrow y:=y+1 \;[]\; x<y \rightarrow x:=x+1 \qquad (*)$$

is miraculous in states satisfying $x=y$, and no implementation exists for such states. One may be tempted to take the Pascal view that (*) could be implemented in the event that $x=y$ by doing nothing, but that's not the semantics we want — if the specification gives no options then it means just that, it is not even giving the option of doing nothing. Asking an implementor to implement a specification when it is miraculous is like asking a person to choose a sweet from an empty bag — it can't be done. If we really want no action in (*) when $x=y$ then we must say so explicitly, as follows

$$x>y \rightarrow y:=y+1 \;[]\; x<y \rightarrow x:=x+1 \;[]\; x=y \rightarrow skip$$

where **skip** (which we shall describe more precisely later) means "do nothing". Although final specifications will never (or should never) be miraculous, in fact miraculous specifications are very useful in the systematic construction of large specifications: we often make small specifications that happen to be miraculous, but which will later be combined with others to form a non-miraculous whole. We will see examples of this shortly.

The module of Figure 2 is still not quite how we would make it in practice: the

**library_catalogue module [BOOK, TITLE, AUTHOR, SUBJECT] =**

>  *books*: set of BOOK;
>  *title*: BOOK $\twoheadrightarrow$ TITLE | books $\subseteq$ dom.title;
>  *authors*: BOOK $\twoheadrightarrow$ sequence of AUTHOR | books $\subseteq$ dom.authors;
>  *subject*: BOOK $\twoheadrightarrow$ SUBJECT | books $\subseteq$ dom.subject;
>
>  **addbook0 =**
>  >  *b*: BOOK, *t*: TITLE, *as*: sequence of AUTHOR, *s*: SUBJECT.
>  >  $\neg$catalogued $\rightarrow$
>  >  >  books:$\cup$ {b} || title:$\oplus$ {(b,t)} || authors:$\oplus$ {(b,as)} ||
>  >  >  subject:$\oplus$ {(b,s)} •
>
>  **catalogued =**
>  >  *b*: BOOK. b$\in$ books •
>
>  **addbook =**
>  >  $\sigma$ addbook0. addbook0 [] oldbook •
>
>  **oldbook =**
>  >  *b*: BOOK.
>  >  catalogued $\rightarrow$ message('Non-unique book identifier') •
>
>  **removebook0 =**
>  >  *b*: BOOK.
>  >  catalogued $\rightarrow$ books:- {b} •
>
>  **removebook =**
>  >  $\sigma$ removebook0.
>  >  removebook0 [] badbook •
>
>  **badbook =**
>  >  *b*: BOOK.
>  >  $\neg$ catalogued $\rightarrow$ message('Unknown book') •
>
>  **booksby =**
>  >  *a*: AUTHOR. {b: b$\in$ books $\wedge$ a$\in$ authors.b} •
>
>  **bookson =**
>  >  *s*: SUBJECT. {b: b$\in$ books $\wedge$ s=subject.b} •

**end**

Fig. 3: Library catalogue version 3

specification we would probably write is shown in Figure 3, with explanations following. The main change that Figure 3 introduces is a stylistic one: the constituent specifications are smaller. We prefer to make lots of small specifications because it makes the task more manageable, and because it increases the chances of reusing old specifications.

In general, the constituent specifications of a module will refer freely to the variables of the module, but they may also refer to other objects. For each named operation we give arbitrary names to such extra objects by means of a piece of text that precedes the definition proper; this text is called the parameter list in the jargon of programming languages but we will use the term "signature". We let σs denote the signature of operation s. For example, in

> AMtime =
> $h$: INTEGER, $m$: INTEGER.
> 0≤h<12 ∧ 0≤m<60 •

> PMtime =
> σAMtime.
> 12≤h<24 ∧ 0≤m<60 •

we could equally well have written $h$: INTEGER, $m$: INTEGER instead of σAMtime.

The signature is just a local extension of the name space, the actual objects associated with the names being left open; we "bind", i.e. attach objects to the names, when we use the operation. Parameter names may be bound in three ways. The first way is the familiar one of supplying an actual parameter, or argument. The meaning of a procedure or function applied to an argument is the body of the procedure or function (i.e. the right hand side of its definition with the parameter list stripped away) with the parameter replaced everywhere with the argument. For example, given

> goodtime =
> $h$: INTEGER, $m$: INTEGER.
> 0≤h<24 ∧ 0≤m<60 •

then goodtime.(12, 15) is equivalent to 0≤12<24 ∧ 0≤15<60, i.e. True. The second way we can bind parameters is by quantification, and we shall see examples of that later. The final way is by implicit binding. An argument may be omitted when there is a similarly named and typed object in scope at the point of "invocation"; the similarly named object is taken as the argument. In other words, the meaning of a procedure or function when "invoked" without arguments is just its body. For example, in

> minutes_since_midnight =
> $h$: INTEGER, $m$: INTEGER.
> goodtime → h*60+m
> [] ¬goodtime → -1 •

goodtime is "invoked" without actual arguments, so the meaning of the invocation is got by substituting its body, giving

> minutes_since_midnight =
> $h$: INTEGER, $m$: INTEGER.
> 0≤h<24 ∧ 0≤m<60 → h*60+m
> [] ¬ 0≤h<24 ∧ 0≤m<60 → -1 •

We could equally well have written

>   minutes_since_midnight =
>     $h$: INTEGER, $m$: INTEGER.
>       goodtime.(h,m) → h*60+m
>     [] ¬goodtime.(h,m) → -1 •

but that takes a little extra writing. We have used implicit binding in Figure 3 when applying **catalogued** within **addbook0**, and again when applying **addbook0** and **oldbook** within **addbook**. **addbook**, for example, is equivalent to — and could have been defined as:

>   **addbook** =
>     $b$: BOOK, $t$: TITLE, $as$: sequence of AUTHOR, $s$: SUBJECT.
>       ¬ b∈ books → books:∪ {b} || title:⊕ {(b,t)} || authors:⊕ {(b,as)} ||
>         subject:⊕ {(b,s)}
>     [] b∈ books → message('Non-unique book identifier') •

Of course, implicit biding is only legitimate when it is used in a context that includes objects whose name and type agree with the signature of the parametrised specification.

Note that parametrisation has low binding power, so the brackets in $x$:TYPEX. (s [] t) are superfluous. Observe, also, that **oldbook** is a partial specification, and that that's perfectly acceptable in the context.

## 3. A closer look at notation

The most important statement that the specification language has over its constituent programming language is what's called the "nondeterministic assignment". This has the form lv:~ P for lv a list of variables and P an assertion, and the following semantics: assign values to the variables in lv so that P holds after the assignment. For example x:~ x=0 asks for x to be given the value 0, and x:~ x>0 asks for x to be given any positive value. We may use primed variables in the P of lv:~ P to denote the values of variables before the assignment: for example, x,y:~ (x<y ∧ y>y') asks for the value of y to be increased and x to be given a value less than the new value of y. Actually we don't make much use of the nondeterministic assignment in this form in the present paper, because we can get by with a simple form of it. The simple assignment x:= e is an abbreviation for x:~ x=e' where e' denotes e with all its specification variables primed. For example, x:= x+1 is equivalent to x:~ x=x'+1, i.e give x a value equal to its old value plus one. The specification x:~ False (where x is any variable or list of variables, even the empty list ε) is legitimate but everywhere miraculous and so wholly unimplementable: it specifies the empty set of programs and is called **miracle**. The specification x:~ True means "give x any value you like", and of course can be implemented trivially.

The parallel assignment x:~P || y:~Q is defined to be equivalent to x,y:~ P∧Q. For example, x:=y || y:=x is equivalent to x:~x=y' || y:~y=x' which is equivalent to x,y:~x=y' ∧ y=x', i.e. exchange the values of x and y. || is evidently commutative and associative, and has identity element ε:~True. ε:~True specifies "do nothing" and is given the name **skip**. Note that in x:~P there is never any need to prime variables in P that do not occur in x; for example x:=y+z || y:~ y>0 is equivalent to x,y:~ (x=y'+z ∧ y>0) — there is no need to prime z because its value is not changed by the assignment. The arguments of || are usually assignments but in fact may be any semicolon-free specification. As a small exercise, consider the specification x:=1 || x:=2 (below, and elsewhere, we intersperse proof steps with their justification enclosed by double quotes):

    x:=1 ‖ x:=2
="definition of simple assignment"
    x:~x=1 ‖ x:~x=2
="definition of ‖ "
    x:~ x=1 ∧ x=2
="logic"
    x:~ False
="definition"
    **miracle**

— x:=1 ‖ x:=2 is a legitimate specification, but it is impossible to implement.

For P an assertion and s a specification, the specification P → s is equivalent to s if the initial state satisfies P, and otherwise it is equivalent to **miracle**. We give ‖ an operator precedence above → . Some laws are:

(A) False → s   =   miracle      for any s
(B) True → s   =   s
(C) P → miracle   =   miracle
(D) (P → s) ‖ t   =   P → s ‖ t
(E) P → (Q → s)   =   P ∧ Q → s

Such laws may seem trivial and even useless, but they are as fundamental to manipulating specifications as the law x+0 = x is to doing arithmetic.

For s and t any specifications, s [] t is a specification meaning "either s or t arbitrarily". For example x:=0 [] x:=1 specifies "let x have the value either 0 or 1". We give [] an operator precedence below → ; so the brackets in ((R → (x:~P ‖ y:~Q )) [] s) are superfluous. It should be intuitively obvious that [] is commutative and associative, and has **miracle** as its identity element. Note that the degree of nondeterminacy is state-dependent. For example, x:=1 [] y≠0 → x:=0 offers the choice of giving x the value 0 or 1 in the case that y differs from 0, and otherwise only the choice 1. The specification P → s [] ¬P → t is equivalent to s in the case that P holds, and otherwise t — so it is similar in style to the Pascal-style **if** P **then** s **else** t. But P → s is not similar to **if** P **then** s — the sense of the latter is captured by P → s [] ¬P → **skip** which is not quite the same as P → s.

The invariant of a module is not simply a comment expressing a good intention, but rather shapes the semantics of the module. Without going into technicalities, the semantics are such that any action that would violate the invariant turns out to be miraculous and so the rules of refinement will never lead to an implementation — there just isn't one. In effect, we have a proof obligation to show that the invariant is maintained, but this is usually just clerical routine. In the case of Figure 3, a cursory inspection of the text suffices to convince that the invariant is maintained: Consider the constraint books ⊆ dom.title, for example. It could only be violated by adding a new book or discarding a title; but when a new book is added then *title* is correspondingly enlarged, while on the other hand the domain of *title* is never decreased, so the constraint is maintained.

## 4. An extended example

We will further illustrate the specification language and style by pursuing the library example further. We are asked to specify a library system in which we may add and remove a book from the library, list the books in a particular subject area or by a particular author, add or remove a library user, enquire about the personal details of a particular user, lend out a book and accept its return, enquire as to whether a particular book is available for loan, and list all the books on loan to a particular user. Moreover, certain operations such as adding a new user to the library are to be restricted to library staff only.

The first step in addressing a specification problem is to isolate good abstractions. This not only helps us to manage the complexity of detail that always threatens to overwhelm us when we tackle a computational problem, but also increases the chances that we'll be able to exploit existing specifications and their implementations, and further that any specification we make might be reusable in the future. It also helps to contain the damage when the requirements are changed. The library system on first inspection appears to comprise three more or less independent components: a catalogue of books, a register of users including staff, and a record of books borrowed. It seems a good idea to model these components in isolation, and then look at their composition. We shall have to be quite determined to treat these as independently as possible, for it is easy to be persuaded that they are inextricably linked together. It would seem difficult to describe the removal of a user from the register, for example, without considering whether the user has library books in his possession, and for that we need to know about the list of borrowings, whereas on the other hand it would seem that we can't describe borrowings without knowing about the register of users! Nonetheless, we shall try to isolate the various concerns.

Let us begin by describing the catalogue of books. That's easy in this case because we have an old specification that we can reuse, the one in Figure 3. But what about protection, and such problems as deleting a book that is on loan? As far as protection is concerned, that's an issue that concerns the library as a whole, and so this is not the level at which to consider it. Neither at this level should we worry about borrowed books: we should just describe the essence of a catalogue, and the specification of Figure 3 does just that.

Next we'll describe the register of users. That's very like a catalogue of books, so we can just rework the specification of Figure 3, arriving at the specification in Figure 4. We use $\delta$ in **enquireuser()** to indicate output parameters (akin to **var** as used in Pascal parameter lists). That the invariant is maintained is evident from a brief inspection of the text.

To describe the list of borrowings we will need to know about calendars so that we can calculate the date on which a book being borrowed is due back. Presumably a calendar module is to be found in the library of specifications we will have built up; an outline of one is given in Figure 5.

The borrowings file is given in Figure 6, with some explanations following.

The **include** clause is equivalent to a textual expansion of the module it names; we have been careful to keep all names distinct to avoid discussing issues of scope and visibility.

Implicit binding is used for **loansto** in the definition of **within**; #loansto is just #{b: (b,p)∈ has}. The constraint of *has* uses binding by quantification (as well as binding by application to an argument). For any functional specification f and any quantifier $Q$ appropriate to the type of f, we may write $Qf$ to denote f with the names in its parameters bound by $Q$. For example, ∃**registered**, where **registered** is defined in Figure 4, denotes ∃p: PERSON. p∈ users, i.e. that at least one person is a registered borrower. This device is used in the constraint of the partial function *has*: ∀(within.limit) is equivalent to ∀p:PERSON. #loansto ≤ limit, and substituting for **loansto** this is equivalent to library_users module [PERSON, NAME, ADDRESS, STATUS] =

>   *users*: set of PERSON;
>   *name*: PERSON ↠ NAME | users ⊆ dom.name;
>   *address*: PERSON ↠ ADDRESS | users ⊆ dom.address;
>   *status*: PERSON ↠ STATUS | users ⊆ dom.status;
>
>   **adduser0** =
>   > *p*: PERSON, *n*: NAME, *a*: ADDRESS, *s*: STATUS.
>   > ¬ registered →
>   > > users:∪ {p} ∥ name:⊕ {(p,n)} ∥ address:⊕ {(p,a)} ∥
>   > > status:⊕ {(p,s)} •
>
>   **registered** =
>   > *p*: PERSON. p∈ users •
>
>   **adduser** =
>   > σ adduser0. adduser0 [] olduser •
>
>   **olduser** =
>   > *p*: PERSON.
>   > registered → message('Person already in library') •
>
>   **removeuser0** =
>   > *p*: PERSON.
>   > registered → users:- {p} •
>
>   **removeuser** =
>   > σ removeuser0. removeuser0 [] baduser •
>
>   **baduser** =
>   > *p*: PERSON.
>   > ¬ registered → message('Not a registered user') •
>
>   **enquireuser0** =
>   > *p*: PERSON; δ *n*: NAME, *a*: ADDRESS, *s*: STATUS.
>   > registered → n:= name.p ∥ a:= address.p ∥ s:= status.p •
>
>   **enquireuser** =
>   > σ enquireuser0. enquireuser0 [] baduser •

end

Figure 4. Register of library staff and borrowers.

∀p:PERSON. #{b: (b,p)∈ has} ≤ limit, i.e. that no person has more than *limit* books on loan.

The operation **ds** (in **return0**) stands for so-called "domain subtraction": has **ds** {b} equals *has* with all pairs whose left hand element is b deleted (there will be precisely one such pair in the present case). Note that the definition of **borrow** has a genuine nondeterminism in that an attempt by a user to borrow an unavailable book when he has calendar module =

    DATE = ...;  "e.g. yymmdd where yy=year, mm=month, dd=day"

    *today*: DATE | validdate.today;

    **validdate** =
        *d*: DATE. ..."d is a legitimate date" •

    **setdate** =
        *d*: DATE.
          validdate → today:= d
        [] baddate •

    **baddate** =
        *d*: DATE.
          ¬validdate → ... •

    **future** =
        *n*: NATURAL. ..."the date n days following today" •

    **update** =
        today:= future.1 •

    ·
    ·
    ·

**end**

Figure 5: A calendar module.

already borrowed to his limit produces one of two error messages, but we don't know which one and we have chosen not to care. Of course, if we really did care then we would specify differently. Again, it is easy to check that the invariant is maintained.

    It only remains to assemble the three pieces. Assembling such components begins with the question "what relationship between the variables do we wish to impose"; when we have answered that, the extra specifying to be done will follow routinely. The catalogue and users modules are independent of one another. The borrowings module is related to the catalogue module in that the books borrowed must be in the catalogue, i.e. dom.has ⊆ books. The relationship between the borrowings module and the users module is simply that the persons to whom books are lent must be registered users, i.e. rng.has ⊆ users. In summary, the extra invariant is

    dom.has ⊆ books ∧ rng.has ⊆ users

Now we just routinely go through the three pieces and ensure that the new invariant is maintained, suitably modifying any action that would violate it.

    First we introduce a new convention: we allow more than one operation in a module to have the same name. When a module contains more than one operation with the same name **n** we need a rule to determine which **n** is being referred to whenever **n** is "invoked". We distinguish two cases: a reference to **n** in an operation whose name is **n**, and library_borrowings module [BOOK, PERSON] =

    include calendar;

    *time*: NATURAL;       "number of days that a user may retain a book"
    *limit*: NATURAL;      "maximum books a user can have on loan"
    *has*: BOOK $\twoheadrightarrow$ PERSON | $\forall$(within.limit);
                           "which books are on loan, and to whom,..."
    *due*: BOOK $\twoheadrightarrow$ DATE | dom.has $\subseteq$ dom.due;   "... and when due back"

    **within** =
        $n$: NATURAL.
            $p$: PERSON. #loansto $\leq n$ •

    **loansto** =
        $p$: PERSON. {$b$: ($b,p$)$\in$ has} •

    **borrow0** =
        $p$: PERSON, $b$: BOOK.
           available $\wedge$ within.(limit-1) $\rightarrow$
                has:$\oplus$ {($b,p$)} || due:$\oplus$ {($b$,future.time)} •

    **available** =
        $b$: BOOK. $b \notin$ dom.has •

    **borrow** =
        $\sigma$ borrow0. borrow0 [] bookout [] nomore •

    **bookout** =
        $b$: BOOK. $\neg$ available $\rightarrow$ message('Book out on loan') •

    **nomore** =
        $p$: PERSON.
           $\neg$ within.(limit-1) $\rightarrow$ message('At limit of borrowings') •

    **return0** =
        $b$: BOOK. $\neg$ available $\rightarrow$ has:ds {$b$} •

    **return** =
        $\sigma$ return0. return0 [] notout •

    **notout** =
        $b$: BOOK.
           available $\rightarrow$ message('Book not on loan') •

end

Fig. 6: Borrowings.

all other instances. In the definition of the operations named n a reference to n is a reference to the n most recently defined in the preceding text. Otherwise, any reference to n refers to the n that occurs last in the text. This renaming convention can save us the trouble of inventing new names; for example, in Figure 6 the successive definitions

>       return0 =
>           $b$: BOOK. $\neg$ available $\to$ has:ds {b} •
>
>       return =
>           $\sigma$ return0. return0 [] notout •

could equally well have been written as

>       return =
>           $b$: book. $\neg$ available $\to$ has:ds {b} •
>
>       return =
>           $\sigma$ return. return [] notout •

Here is a more algorithmic formulation of the rule to resolve references to a name **n** which is defined more than once: attach subscripts 0, 1, 2, ... to each defining occurrence of **n** in order of appearance in the text; in the body of each $n_i$ attach subscript i-1 to each **n** therein; and finally drop the subscript from the last defining occurrence of **n**. Applying this algorithm to the successive definitions of **return** above yields

>       $return_0$ =
>           $b$: BOOK. $\neg$ available $\to$ has:ds {b} •
>
>       return =
>           $\sigma$ $return_0$. $return_0$ [] notout •

which is pretty much the definitions that appear in Figure 6. The main advantage of this convention is that it gives us a mechanism for changing the definition of an operation without changing its name; if we could not retain the old name then we would have the bother of changing every existing reference to the old name. Some care is needed when reusing a name, because reuse precludes any further reference to the old definition.

We proceed to inspect **library_catalogue** for violations of dom.has $\subseteq$ books; this could only be violated by an action that removes an element from *books*, and that can only occur in **removebook0**, which indeed needs some revision:

**removebook0** =
>       $\sigma$ removebook0.
>           available $\to$ removebook0 [] bookout •

There is no need to expand such definitions, but it may be helpful to see just one:

   **removebook**
="definition from Figure 3"
    $\sigma$ removebook0. removebook0 [] badbook
="definition of removebook0 above"
    $\sigma$ removebook0. (available $\to$ removebook0 [] bookout) [] badbook
="[] associative"
    $\sigma$ removebook0. available $\to$ removebook0 [] bookout [] badbook
="definition of removebook0 from Figure 3"

```
        b: BOOK.
            available → (catalogued → books:- {b})
        [] bookout [] badbook
="(E) from section 3"
        b: BOOK.
            available ∧ catalogued → books:- {b}
        [] bookout [] badbook
="definitions"
        b: BOOK.
            b∉ dom.has ∧ b∈ books →  books:- {b}
        [] b∉ books → message('Unknown book')
        [] b∈ dom.has → message('Book out on loan')
```

Going through the same procedure for **library_users** we find that we have to enlarge **removeuser0**, and checking through **library_borrowings** we find that action **borrow0** needs modifying.

Finally, it is at this level that we should deal with querying the status of a book; it could not be done sooner as it involves both the catalogue of books and the borrowings file. The combined specification is shown in Figure 7. Note that we have composed the module of Figure 7 with pieces that are not as small as we have employed in earlier modules, just to show an alternative style.

Now we might proceed to design a protected library system in which certain operations are restricted to staff, but we shall leave the library there.

## 5. Discussion of the specification

One half of a specification methodology is the systematic construction of specifications, and we tried to show in the previous section that the methodology we are employing is capable of going some way towards achieving this. We have avoided many issues however: small issues such as initialising variables and exporting names, and larger issues such as scope of declarations, and parametrisation of modules. The usefulness of parametrisation is hinted at by our use of unspecified types, and further by the evident similarity between the modules of Figures 3 and 4, which we did not exploit. Despite the issues dodged, we feel that the specification style we have been illustrating suits its purpose well, but it would be wrong to claim too much on the evidence of one example. Let us consider why we specified the library as we have, and how we might have done it differently, both for better and for worse.

At first sight it may be irritating to have to read modules composed of so many small pieces: the module of Figure 3, for example, seems rather fragmented in comparison to the functionally equivalent one of Figure 2. But there are good reasons for keeping the pieces small. First of all, that's how we make them. Secondly, it increases the opportunities for reusing components, and indeed the final module makes use of operations **catalogued** and **available** from Figure 3, operations that do not exist in their own right in the catalogue of Figure 2. Thirdly, modules composed of small pieces make it easier to adapt the module to changing requirements. Suppose, for example, that we've just been told that a new feature is to be added to the catalogue module: all operations that change the state are to be "undoable", i.e. an operation **undo** is to be added that enables the user to recover the state that existed immediately prior to the current one. We can quite nicely build an extension on the catalogue of Figure 3 — see Figure 8 — but Figure 2 offers no such possibility: we would

**library_system module** [BOOK, TITLE, AUTHOR, SUBJECT,
                                PERSON, NAME, ADDRESS, STATUS] =

   **include** library_catalogue[BOOK, TITLE, AUTHOR, SUBJECT],
              library_users[PERSON, NAME, ADDRESS, STATUS],
              library_borrowings[BOOK, PERSON];

   | dom.has $\subseteq$ books $\wedge$ rng.has $\subseteq$ users;

   **removebook0** =
          $\sigma$ removebook0.
              available $\rightarrow$ removebook0 [] bookout •

   **removeuser0** =
          $\sigma$ removeuser0.
              $\neg$ hasbook $\rightarrow$ removeuser0 [] bookdue •

   **hasbook** =
          $p$: PERSON. $p \in$ rng.has •

   **bookdue** =
          hasbook $\rightarrow$ message('Has a book on loan') •

   **borrow0** =
          $\sigma$ borrow0.
              registered $\wedge$ catalogued $\rightarrow$ borrow0
          [] baduser [] badbook •

   **bookonshelf** =
          $b$: BOOK. catalogued $\wedge$ available •

**end**

Figure 7: The library system

have to decompose it to its constituent parts, make the change, and reassemble. Note that although the specification of Figure 8 retains a complete copy of the old state, the semantics do not oblige an implementation to follow suit, and of course we would derive an implementation that just keeps a note of the difference between the present state and its predecessor.

The specifier's slogan is "Be minimalist!". The less said in a specification the better, not only because it increases the options when we come to implementing, but also because once we have set down the framework of a specification, much of the detail follows almost mechanically. Let's examine the library specification to see where we might have been tempted to say more, or could have done with saying less.

Consider the catalogue of Figure 3. Thinking as programmers, ever on the lookout to save space, we might have been tempted to to specify the body of **removebook0** as

   catalogued $\rightarrow$ books:- {b} || title:ds {b} || authors:ds {b} || subject:ds {b}

There is nothing wrong with the new definition, but it adds pointlessly to the text and so just

**undoable_library_catalogue module** [BOOK, TITLE, AUTHOR, SUBJECT] =

    **include** library_catalogue[BOOK, TITLE, AUTHOR, SUBJECT];

    *booksbefore:* set of BOOK;
    *titlebefore:* BOOK $\twoheadrightarrow$ TITLE;
    *authorsbefore:* BOOK $\twoheadrightarrow$ set of AUTHOR;
    *subjectbefore:* BOOK $\twoheadrightarrow$ SUBJECT;

    **save** =
        booksbefore:= books || titlebefore:= title ||
        authorsbefore:= authors || subjectbefore:= subject •

    **addbook0** =
        σ addbook0. addbook0 || save •

    **removebook0** =
        σ removebook0. removebook0 || save •

    **undo** =
        books:= booksbefore || title:= titlebefore ||
        authors:= authorsbefore || subject:= subjectbefore || save •

**end**

Fig. 8: Library catalogue with undo command.

increases the work to be done in discharging proof obligations. Functionally the specifications are identical and admit precisely the same implementations.

There is only one guard in Figure 3, namely **catalogued** (and its negation). Employing this guard was a design decision in that the specification without it would still have been consistent with the informal requirements, although we would now have a different specification. We employed **catalogued** because we surmised that an attempt to add a book that was already in the library is indicative of something amiss with the world, but we could equally well have taken the view that all was in order, perhaps that the librarian was simply changing the attributes of the book. It is an advantage of formal specifications that they invite us to make decisions early, when the options are still open.

Turning to the borrowings module of Figure 6, consider **borrow0**. Its guard is a conjunction of two conditions, **available** and within.(limit-1). The first represents a design decision and could well have been omitted — one can easily think of a situation where it is reasonable to borrow a book that is on loan. The second condition, however, is motivated by the need to maintain the invariance of $\forall$(within.limit). Yet it turns out, and this may be surprising, that it too is dispensable — the functionality of **borrow0** is not changed a whit by its presence or absence. The reasons for this are a consequence of the formal semantics that we attach to assignments in the presence of module invariants. Of course, we will have to supply the guard at some stage in the derivation of an implementation, but the beauty of the refinement calculus is that the guard can be *calculated*, almost automatically from the statement it is to guard and the invariant that is to be maintained. Had we omitted the condition, we would inevitably meet it anyhow. The module of Figure 7, also contains a good deal of redundancy, and most of it could in fact have been calculated once the invariant was decided upon. (We could significantly shorten the description of modules — while imposing extra calculational work on the implementor — by the use of notational devices that exploit these redundancies, but we shall not do so in the present paper.) Having made the point that specifications can leave out a great deal of what we might have regarded as crucial, perhaps we should also make the point that redundancies may be advantageous in so far as they may improve clarity.

## 6. A summary of refinement

The other half of the methodology is the transformation of specifications into programs, which we now address. Suppose we want to make a Pascal program that sorts array $b$, indexed from 0 to N-1, in situ; we may specify the problem as that of implementing "b:= sorted.b" where sorted yields the lexicographically least permutation of its argument. We might then proceed to write a first outline thus:

```
i:= 0;
while i<> N do begin
        "k:= index of least value in b[i], b[i+1], ..., b[N-1]";
        "swap b[i], b[k]";
        i:= i+1
end
```

Observe that the texts within quotes are informal specification statements. The above in formal dress is

```
i:= 0;
while i<> N do begin
        k:~ i≤k<N ∧ b[k]=min.b[i..N-1];
        b[i]:= b[k] ∥ b[k]:= b[i];
        i:= i+1
end
```

Next we would proceed to tackle "k:~ i≤k<N ∧ b[k]=min.b[i..N-1]" on its own, and then substitute the solution into its place in the body of the loop, and so on. So the specification language also serves as a fancy pseudo-code, that supports the gradual translation of the original specification into a program. This style of programming is what is familiarly known as stepwise refinement; in formal dress it is the application of what we call the "refinement calculus".

A specification t is said to be a "refinement" of specification s, informally speaking, whenever a customer asking for s would be willing to accept t. The two specifications need not be quite functionally equivalent because the refining specification may discard choices offered by the original specification. We will write s ≤ t to denote "s is refined by t". For example,

```
b:= sorted.b            ≤           i:= 0;
                                    while i<> N do begin
                                            k:~ i≤k<N ∧ b[k] = min.b[i..N-1];
                                            b[i]:= b[k] ∥ b[k]:= b[i];
                                            i:= i+1
                                    end
```

When $s \leq t$ and $t$ is a program we say that $t$ "implements" s. By a program we mean a specification that does not contain certain constructs that we deem to be non-algorithmic because they are impossible or too expensive to implement; there is some flexibility here to allow for the ingenuity of the compiler-writer, but we can just think of Pascal-like programs. Note that [] and $\rightarrow$ are part of the programming language, and so is || when its arguments are simple assignments, although we may prefer to refine it away on efficiency grounds. Here are examples of the refinement relation (the variables are of type INTEGER):

(i)     $x:\sim x>0$          $\leq$    $x:= 15$
(ii)    $x:= |x|$             $\leq$    $x\leq 0 \rightarrow x:= -x$ [] $x\geq 0 \rightarrow$ skip
(iii)   $x:=y \,||\, y:=x$    $\leq$    (z: INTEGER; $z:= x; x:= y; y:= z$)
(iv)    $x:=0$ [] $x:= 1$     $\leq$    $x:= 1$
(v)     $x:= 3 \,||\, y:= x*x$ $\leq$   $y:= x*x; x:= 3$

The refinement relation is transitive; this is important because it means that we don't have to implement a specification in one great action, but can proceed from specification to program through a series of small steps and be sure that what we end up with indeed refines the original specification.

If we give a formal semantics to specifications, and we can, then we can formally define refinement and hence derive a collection of useful laws. Among such laws are

(F)     s              $\leq$    $P \rightarrow s$ [] $\neg P \rightarrow s$
(G)     $Q \rightarrow x:\sim Q$  $\leq$    $Q \rightarrow$ skip

— (F) allows us to proceed by case analysis, and (G) allows us to replace a non-algorithmic construct with an algorithmic one. The refinement calculus is a collection of such laws. Rather than guess a refinement or implementation and then verify it afterwards, the motivation behind a calculus is that we transform the specification into a program by systematically applying the laws to its components until all the non-algorithmic or unacceptably inefficient constructs have been eliminated. The maker of the laws needs a formal semantics to justify them, but the semantics play no role in their application — for that one just needs to know the laws of logic and the problem domain.

A very desirable property of refinement gives us the ability to refine a specification by refining its components in relative isolation, and then gluing the refinements together with the same structure as the original specification. For example, we should be able to infer from the examples above without further work that

(vi)    $x:\sim x>0; (x:=y \,||\, y:=x)$    $\leq$   $x:= 15; (x:=y \,||\, y:=x)$
(vii)   $x:\sim x>0$ [] $x:=y \,||\, y:=x$    $\leq$
        $x:= 15$ [] (z: INTEGER; $z:= x; x:= y; y:= z$)

— refinement (vi) follows from (i) and the fact that semicolon happens to be what we may call "refinement-preserving" in its left argument, and (vii) follows from (i) and (iii) and the fact that choice happens to be refinement-preserving in both its arguments (in fact, semicolon is refinement-preserving in its right argument also). We prefer the shorter word "monotonic" to "refinement-preserving". Formally, an operation $F(x)$ on specifications is monotonic if for all specifications s and t, $F(s) \leq F(t)$ whenever $s \leq t$. So the statement "semicolon is monotonic in its left argument" means that $s; u \leq t; u$ whenever $s \leq t$. An implementation can preserve the structure of its specification just to the extent that its combinators are monotonic. We have already argued that structure-preserving implementations are a good thing, so are the combinators we have been using monotonic? The good news is that choice, semicolon, $\rightarrow$ , and parametrisation are monotonic, but parallel composition is not. (The left argument of $\rightarrow$ is an expression, and for expressions refinement is just functional equality.) We will consider the implications of this shortly.

So far we have been talking about what is called "procedural refinement". We will also want to replace some variables with more efficiently implementable ones, and make whatever changes are thereby induced on the statements that act on these variables; this kind of refinement is called "data refinement". For example, in implementing the library catalogue we may want to replace *books* of type set of BOOK with *books1* of type sequence of BOOK, because sets are not part of the programming language. A consequence is that the assignment books:= books $\cup$ {b} must be transformed to, perhaps, the assignment books1:= books1 + ⟨b⟩ where + denotes sequence concatenation and ⟨b⟩ denotes the sequence containing b only. There is a calculus of data refinement that allows us to make these transformations constructively, by calculation. The calculus is based on a relation $\lll_I$ on specifications that may operate on different variables, I being some predicate — called the "abstraction invariant", that relates the two sets of variables. s $\lll_I$ t is defined to hold, roughly speaking, whenever the effect of s operating on one set of variables is mimicked by t operating on the other, where I states the nature of the mimicking. For example, we can derive

books:= books $\cup$ {b}    $\lll_{I0}$    books1:= books1 + ⟨b⟩

where I0 is the relation "*books* = the set of values in sequence *books1*". There will usually be several candidates for an abstraction invariant. For example, when replacing *books* with *books1* we might have chosen the stronger predicate I1 as follows: "*books* = the set of values in sequence *books1*, and the values in *books1* are distinct". This would lead to a somewhat different translation of the statements that refer to *books*, for example

books:= books $\cup$ {b}    $\lll_{I1}$    b $\in$ books1 $\rightarrow$ skip
                                            [] $\neg$ b $\in$ books1 $\rightarrow$ books1:= books1 + ⟨b⟩

In any case, once we have decided on the new variables and fixed on the abstraction invariant, the translation of the specification to a similar one operating on the new variables is by a systematic application of the rules of the calculus. The rules are either statements about monotonicity, or they reduce data refinement to procedural refinement; an example of such a rule is

(H) s; t $\lll_I$ u; v whenever s $\lll_I$ u and t $\lll_I$ v

— this says we can data refine a composition by data refining its components in isolation.

We can extend the notion of refinement to modules M1 and M2 when the "interface" of M1 is a subset of the interface of M2. By the interface of a module we mean the set of pairs ⟨p, TP⟩ for each operation name p in the module, where TP denotes the types of its parameters (with an indication of which are output parameters) and result (if any). Refinement among modules does not require that the variables of one module have anything in common with the variables of the other. In fact, there are situations where one wishes to relax the requirement that a refining module share a common interface with its parent, but we will not admit that generality in the present treatment. When modules M1 and M2 share a common interface *and* a common set of variables then M1 $\leq$ M2 holds just when each operation of M1 is refined by the similarly named operation of M2; this is just as we would intuitively expect. More generally, if the interface of M2 is a superset of the interface of M1 but it uses a possibly different set of module variables, then M1 $\leq$ M2 holds just when there is a predicate I (the abstraction invariant) relating the two sets of variables such that each operation of M1 is related to the similarly named operation of M2 by «$_I$. In practice we do not prove that one module refines another, but refine modules constructively; we decide on the new set of variables and an abstraction invariant, and then make the new module by operating on the old module with the laws of data and procedural refinement.

For a more comprehensive treatment of the refinement calculus see [1-6].

## 7. Implementing the library

One way to implement the library described by the specification of Figure 7 is to expand textually all the definitions and **include**'s, arriving at a module in the style of Figure 2, and then implement the procedures and functions. But that is not what we want because, for reasons given in the introduction, we would like to give each module an independent existence, as far as we can.

We will give an informal outline of how we might go about implementing each module, beginning with the catalogue of Figure 3. Examining the module constraints, we observe that they confer some freedom on the implementor. For example, we can maintain the constraint on *title* while strengthening it to books = dom.title. This will lead to a more space-efficient implementation, so we decide to pursue it. The theory of data refinement supplies all the laws we need to calculate the new specification. (For interest, the abstraction invariant is title$_{new}$ = title **dr** books where for any partial function f and set s, f **dr** s equals f with those pairs whose first element is not in s removed; the operation **dr** is known as "domain restriction".) The refinement results in what one would expect; **removebook0**, for example, becomes

    **removebook0** =
        *b*: BOOK.
        catalogued → books:- {b} ‖ title:**ds** {b} •

We can strengthen the constraints on *authors* and *subject* similarly. After that, we might choose to get rid of the parallel compositions; it so happens — as will be intuitively evident — that those in Figure 3, as well as those introduced by the strengthening of the invariant, can be replaced with a semicolon.

Examining the parameter and result types of the procedures and functions we see that they are all supplied externally — we are assuming in particular that unbounded sequences are implemented, which is not too unreasonable for the small sequences likely to be involved in the present case. We would have to change the interface of the module if "sequence of" is not a type constructor of the programming language. Next we should consider how to implement the variables. Perhaps we will decide on linked lists of records, one record per book recording its identifier, title, authors, and subject. Having chosen the data structures and abstraction invariant, we apply data refinement to generate the new module. After this, the module will be close to an implementation. In any case, all the combinators are now monotonic, so each little piece can be worked on in isolation. Finally we discard the module invariants. We will have arrived at a module that behaves like that of Figure 3, but can be compiled and executed.

A novel aspect of this implementation style is that the final program will contain procedures whose bodies are partial (i.e. miraculous in some states). When called outside their domain of applicability they do not abort or raise some error condition. Rather they "return" carrying some indication of an attempt at miraculous behaviour, and the system will then look for some other way forward.

Implementing the undoable catalogue of Figure 8 has the extra difficulty of a named operation, **save**, being an argument of the non-monotonic ∥. We would first have to replace ∥, carried out in the case of **removebook0**, for example, thus:

    removebook0
="definition"
    σremovebook0. removebook0 ∥ save
="old definition of removebook0"
    σ removebook0. (catalogued → books:- {b}) ∥ save
="(D) in Section 3"
    σ removebook0. catalogued → books:- {b} ∥ save
="∥ commutative"
    σ removebook0. catalogued → save ∥ books:- {b}
="∥ may be replaced with semicolon when the variables being assigned to in the
   left hand argument do not appear in the right hand argument"
    σ removebook0. catalogued → save; books:- {b}

So we can reuse the old implementation, but we have to "open up" its parent specification. (Note also that we now see it would have been advantageous to have made a named operation of books:- {b} in the specification of Figure 3, and similarly for the body of **addbook0**. We could then have left the replacement of books:- {b} in the above with its implementation to the compiler.) The implementation of the undoable catalogue needs further "inside information" in that the implementation of **save** will have to take account of the implementation of *books* in Figure 3.

The modules of Figures 4 and 6 may be implemented similarly to that of Figure 3. The three modules come together in the library system of Figure 7, whose implementation is now easy, because the modules **include**'d can be replaced by their implementations, without any extra work. The only operation of Figure 7 that has to be implemented is **hasbook**; that can be easily done, but we need to exploit the knowledge of how *has* was implemented when data refining the borrowings module of Figure 6. With hindsight, **hasbook** should probably be moved to the borrowings module, as a companion to **available**.

In summary, we have not quite managed to implement each module independently, but we went a good way towards that goal. What we did achieve was the ability to implement modules in isolation provided we do so in an order consistent with that induced in the obvious way by the **include** clauses. It is pleasing that the old implementations were reusable without change, despite the fact that their constituent operations are altered significantly by the modules that **include** them.

We pay a price for the modular implementation, however. When implementing a module in isolation we can only act on knowledge of the actions of that module alone; we elect not to take into account how other modules interact with the present one. For example, if we implemented the catalogue and borrowings modules in parallel we might choose integrated data structures that would almost certainly be more space-efficient over the system as a whole. Although, we are probably prepared to pay a considerable price for the benefits of the modular implementation, the price may sometimes be unacceptably high, and we may have to restructure the specification, at least partly, in the process of implementation.

## 8. Conclusion

We have outlined a specification methodology that tries to do justice to the dual role of specifications in designing and implementing systems. The method is based on the view that a specification language is an enrichment of a programming language; it admits ease of expression although some of its elements may be unimplementable. The attraction of the approach is that specifications and programs inhabit the same semantic world and so we can easily give a precise meaning to a notion of refinement that supports the implementation of programs by mathematical transformation. We gave a brief outline of this theory, and showed that it could even be applied to construct specifications mechanically once enough of their framework has been set down. The implementation can proceed in small steps, retaining the structure of the parent specification, thanks to the monotonicity of the constructors of the language. The language supports the modular construction and implementation of specifications, with some benefits in the area of reusing specifications and their implementations.

## Acknowledgements

The library example is based on a similar one presented by Steve King and Ib Sørenson at the 1988 York Refinement Workshop. The extensive use of partial functions, and the philosophy of specifying incrementally, is borrowed from Z. Thanks to Moira Norrie and David Harper for very detailed criticisms of an earlier draft, and to Cliff Jones for critical comments.

## References

1. R. J. R. Back, *Correctness Preserving Program Refinements: Proof Theory and Applications*, Tract 131, Mathematisch Centrum, Amsterdam, 1980.

2. R. J. R. Back, A calculus of refinement for program derivations, Report Ser. A 54, Department of Computer Science, Swedish University of Abo (1987).

3. C. C. Morgan, The specification statement, *ACM TOPLAS* **10** (1988) 403-419.

4. C. C. Morgan and P. H. B. Gardiner, Data refinement by calculation, *Acta Informatica*, to appear.

5. J .M. Morris, Programs from specifications, in *Formal Development of Programs and Proofs*, ed. E. W. Dijkstra, Addison-Wesley, Reading (Mass), 1989.

6. J. M. Morris, Laws of data refinement, *Acta Informatica* **26** (1989) 287-308.

7. J. M. Spivey, *Understanding Z*, Cambridge University Press, Cambridge, 1988.

# Sannella

In many cases the extended language in which refinements begin was indeed made by extending a language originally designed for programming. That is not to say it should be that way, but it is to some extent to be expected from the historical development of the subject: computers arrived first, and then their programming languages.

For ML, a functional programming language, the extension has an equational flavour: the specifications state only properties of the ultimate computation, as equalities between functional expressions. Although less familiar to programmers generally, such axiomatic descriptions have in the past attracted substantial interest in their own right; and so there is already a reasonably broad agreement about the way in which they should behave.

The advantage of understanding well the behaviour of specifications, independently of the language into which they are refined, is that it makes easier the decisions about the way in which they may be combined, systematically varied, and reused. With the axiomatic specifications of this paper is that particularly clear.

# Formal program development in Extended ML for the working programmer

Donald Sannella*

### Abstract

Extended ML is a framework for the formal development of programs in the Standard ML programming language from high-level specifications of their required input/output behaviour. It strongly supports the development of modular programs consisting of an interconnected collection of generic and reusable units. The Extended ML framework includes a methodology for formal program development which establishes a number of ways of proceeding from a given specification of a programming task towards a program. Each such step gives rise to one or more proof obligations which must be proved in order to establish the correctness of that step. This paper is intended as a user-oriented summary of the Extended ML language and methodology. Theoretical technicalities are avoided whenever possible, with emphasis placed on the practical aspects of formal program development. An extended example of a complete program development in Extended ML is included.

## 1 Introduction

The ultimate goal of work on program specification is to establish a practical framework for the systematic production of correct programs from requirements specifications via a sequence of verified-correct development steps. Such a framework should ideally have a number of desirable characteristics. Among other important issues are the following:

**Formality:** The outcome of the program development process is guaranteed to be correct with respect to the original requirements specification if development steps are proved correct in some formal logical calculus which is sound with respect to a complete mathematical semantics of the specification and programming languages involved. Any hedge on this strictly formal point of view invalidates all guarantees.

**Methodology:** Given a specification of a programming task, it is helpful if the framework provides some form of direction in working towards a solution. One possibility is if the framework sets forth a certain number of kinds of development steps which apply to specifications of a given form, together with the conditions which must be established in order to guarantee correctness. Coming up with development steps is a difficult creative task and standardized methods of making progress are very important in reducing this difficulty to a manageable level.

**Modularity:** Large programs should be built in a modular fashion from small and relatively independent program units, and the framework should support such an approach. Apart from the advantages which this gives in allowing large programming tasks to be broken into a number of smaller and more manageable separate tasks, this allows previously-developed programs and components of such programs to be reused in other programs. One may imagine formal program development becoming almost practically feasible in spite of its great unitary cost because of the potential of spreading this cost over many different projects.

---
*Laboratory for Foundations of Computer Science, Department of Computer Science, Edinburgh University.

**Machine support:** The eventual practical feasibility of systematic program development by stepwise refinement hinges on the availability of computer-aided tools to support various development activities. This is necessary both because of the sheer amount of (mostly clerical) work involved and because of the need to avoid the possibility of human error.

This leaves aside many questions, including: How may the original requirements specification be formulated so as to ensure that it accurately expresses the needs of the customer? How should choices between some number of possible development paths be made? Such issues are no less important than those mentioned above, but they will not be addressed here.

Extended ML is a framework for the formal development of programs in the Standard ML programming language from high-level specifications of their required input/output behaviour. Extended ML is a completely formal framework with a very extensively-developed mathematical basis in the theory of algebraic specifications. It strongly supports the development of modular programs consisting of an interconnected collection of generic and reusable units. The Extended ML framework includes a methodology for formal program development which establishes a number of ways of proceeding from a given specification of a programming task towards a program. Each such step (modular decomposition, etc.) gives rise to one or more proof obligations which must be proved in order to establish the correctness of that step. On the minus side, at present Extended ML can only be used to develop programs written in a small purely functional subset of Standard ML, and a computer-aided system to support program development is still in the design stage.

The Extended ML language is a wide-spectrum language which encompasses both specifications and executable programs in a single unified framework. It is a simple extension of the Standard ML programming language in which axioms are permitted in module interfaces and in place of code in module bodies. This allows all stages in the development of a program to be expressed in the Extended ML language, from the initial high-level specification to the final program itself and including intermediate stages in which specification and program are intermingled.

Formally developing a program in Extended ML means writing a high-level specification of a generic Standard ML module and then refining this specification top-down by means of a sequence (actually, a tree) of development steps until an executable Standard ML program is obtained. The development has a tree-like structure since one of the ways to proceed from a specification is to decompose it into a number of smaller specifications which can then be independently refined further. In programming terms, this corresponds to implementing a program module by decomposition into a number of independent sub-modules.

This paper is intended as a user-oriented summary of the Extended ML language and methodology. Theoretical technicalities are avoided whenever possible, with emphasis placed on the practical aspects of formal program development. Some of the details (mainly those concerned with *behavioural equivalence*) are glossed over in the interests of simplicity. Section 3 provides an overview of the Extended ML language, preceded by a brief review of the Standard ML programming language on which it is based in Section 2. Section 4 presents a methodology for developing Standard ML programs from Extended ML specifications by stepwise refinement. An extended example of a complete program development in Extended ML is included in Section 5; this is the most important section of the paper for the reader who merely wishes to get a taste of what formal program development is like. Finally, Section 6 concludes with some remarks about some potential areas of future progress. Readers who are interested in the theory which underlies Extended ML should consult [ST 89]; among other things, this explains in detail why the formal program development process outlined here is guaranteed to yield a program which is correct with respect to the original specification of requirements.

## 2 An overview of Standard ML

The aim of this section is to briefly review the main features of the Standard ML programming language which are relevant to Extended ML, in an attempt to make this paper self-contained. A complete description of the language appears in [HMM 86], and a complete formal semantics is in [HMT 89] which also includes historical comments on the development of the language. The features of Standard ML are introduced at a more tutorial level in [Wik 87] (core language only), [Tofte 89] (mainly module language), [Har 89], and [Reade 89].

Standard ML consists of two sub-languages: the Standard ML "core language" and the Standard ML "module language". The core language provides constructs for programming "in the small" by defining a collection of types and values of those types. Programs written in the core language look very similar to programs in Hope [BMS 80], Miranda [BW 88] or Haskell [HW 89]. The module language provides constructs for programming "in the large" by defining and combining a number of self-contained program units. These sub-languages can be viewed as more or less independent since there are relatively few points of contact between the sub-languages. A similar modularization mechanism could be added to other programming languages; see [SW 87] for the design of an ML-style module system for Prolog.

## 2.1 The Standard ML core language

Standard ML is a strongly typed language. Every expression has a type which is inferred automatically by the Standard ML compiler. Expressions are required to obey the typing rules before being evaluated, and a well-typed expression is guaranteed to produce no run-time type errors. A new type is defined by giving its name and listing the ways in which values of that type may be constructed from values of other types. For example, the type of integer sequences may be defined as follows:

```
datatype sequence =
    nil
  | cons of int * sequence
```

(int is the type of integers, which is built-in). A typical value of type sequence is then:

```
cons(3,cons(4,cons(0,cons(3,nil))))
```

Conceptually, every value in Standard ML is represented as a term consisting of a *constructor* applied to a number of sub-terms, each of which in turn represents another value. In the above definition, nil is a nullary constructor and cons is a binary constructor (of type int * sequence -> sequence). Constructor functions are uninterpreted; they just construct. There is no need to define a lower-level representation of sequences in terms of arrays or pointers. Note that type definitions may be recursive, as in the above example. The type of integers may be viewed as if it were defined as follows:

```
datatype int = ... | ~3 | ~2 | ~1 | 0 | 1 | 2 | 3 | ...
```

Other built-in types include booleans (bool) and character strings (string), and there are a few built-in functions such as <= : int * int -> bool (less than or equal), size : string -> int, and not : bool -> bool. The function = : $t * t$ -> bool (and its negation, <>) is automatically provided for any user-defined type $t$.[1] It is possible to give new names to existing types and type expressions:

```
type age = int
```

This creates an additional name age for the existing type int.

Functions are defined by a sequence of one of more equations, each of which specifies the value of the function over some subset of the set of possible argument values. This subset is described by a *pattern* (a term containing constructors and variables only, without repeated variables) on the left-hand side of the equation. The pattern is thereby used for case selection and variable binding. For example:

---
[1]This is the case for the subset of the Standard ML core language used here, but not in general.

```
fun length(nil) = 0
  | length(cons(a,s)) = 1 + length(s)
```

This defines a function `length : sequence -> int` (this type is inferred automatically by the Standard ML compiler). One way of viewing such a definition is as a set of rewrite rules:

```
length(cons(3,cons(4,cons(0,cons(3,nil)))))
                    ⇒   1+length(cons(4,cons(0,cons(3,nil))))
                    ⇒   1+(1+length(cons(0,cons(3,nil))))
                    ⇒   1+(1+(1+length(cons(3,nil))))
                    ⇒   1+(1+(1+(1+length(nil))))
                    ⇒   1+(1+(1+(1+0)))
                    ⇒   4
```

Function definitions are often recursive, as in this example; of course, this defines a terminating function only if the recursion is well-founded in the usual sense. The patterns on the left-hand side of equations should normally be disjoint and should exhaust the possibilities given in the definition of the argument type(s). Constants may also be defined:

```
val three_ones = cons(1,cons(1,cons(1,nil)))
```

Such definitions may not be recursive. Constants and functions are referred to collectively as *values*.

Values may be defined in terms of other values, of course. In the following example, a function to sort a sequence into ascending order is defined using an auxiliary function which inserts an integer into a ordered list:

```
fun insert(a,nil) = cons(a,nil)
  | insert(a,cons(b,s)) = if a<=b then cons(a,cons(b,s))
                          else cons(b,insert(a,s))

fun sort nil = nil
  | sort(cons(a,s)) = insert(a,sort s)
```

This defines `insert : int * sequence -> sequence` and `sort : sequence -> sequence`. Evaluating the expression:

```
sort(cons(11,cons(5,cons(8,nil))))
```

will yield the result:

```
cons(5,cons(8,cons(11,nil))) : sequence
```

The Standard ML core language includes an assortment of other features, but we will only be concerned with simple type definitions and value definitions such as those in the examples above. Features which we will not use include: polymorphic types (it is possible to define sequences of values of type $\alpha$, for arbitrary $\alpha$, and define functions over such types which work for any $\alpha$); higher-order functions and functions as first-class citizens (which means that values can have types like `(int -> int) -> (sequence -> sequence)` and functions can be embedded in data structures); and imperative features (references and exceptions). We restrict ourselves to this simple "pure" subset of the Standard ML core language because adequate (algebraic-style) formal foundations for the additional features are not yet available. To keep things simple we will make the additional assumption throughout this paper that all functions we deal with are total.

## 2.2 The Standard ML module language

The Standard ML module language provides mechanisms which allow large Standard ML programs to be structured into self-contained program units with explicitly-specified interfaces. Under this scheme, interfaces (called *signatures*) and their implementations (called *structures*) are defined separately. Every structure has a signature which gives the names of the types and values defined in the structure. Structures may be built on top of existing structures, so each one is actually a *hierarchy* of structures, and this is reflected in its signature. Components of structures are accessed using qualified names such as A.B.n (referring to the component n of the structure component B of the structure A). *Functors* are "parameterized" structures; the application of a functor to a structure yields a structure. A functor has an input signature describing structures to which it may be applied, and an output signature describing the structure which results from such an application. It is possible, and sometimes necessary to allow interaction between different parts of a program, to declare that certain substructures (or just certain types) in the hierarchy are identical or *shared*. This issue will be discussed later in this section.

An example of a simple modular program in Standard ML is given below. This generalizes the program above for sorting a sequence of integers, by allowing a sequence of values of arbitrary type to be sorted provided an order relation is supplied.

```
signature PO =
    sig
        type elem
        val le : elem * elem -> bool
    end

signature SORT =
    sig
        structure Elements : PO
        datatype sequence =
            nil
          | cons of Elements.elem * sequence
        val sort : sequence -> sequence
    end

functor Sort(X : PO) : SORT =
    struct
        structure Elements = X
        datatype sequence =
            nil
          | cons of Elements.elem * sequence
        fun insert(a,nil) = cons(a,nil)
          | insert(a,cons(b,s)) = if Elements.le(a,b) then cons(a,cons(b,s))
                                  else cons(b,insert(a,s))
        fun sort nil = nil
          | sort(cons(a,s)) = insert(a,sort s)
    end
```

This defines a functor called Sort which may be applied to any structure matching the signature PO, whereupon it will yield a structure matching the signature SORT. In order for the definition of Sort to be correctly typed, the body of Sort must define a structure which contains: a substructure called Elements which matches PO; a type called sequence having constructors called nil and cons with the types given, and no other constructors; and a function called sort with the type given. The definition of Sort is indeed correctly typed, and this is determined automatically at compile time.

We can define a structure of signature PO and apply Sort to this structure as follows:

```
structure IntPO : PO =
   struct
      type elem = int
      val le = op <=
   end

structure SortInt = Sort(IntPO)
```

Now, SortInt.sort may be applied to the sequence

```
SortInt.cons(11,SortInt.cons(5,SortInt.cons(8,SortInt.nil)))
```

to yield

```
SortInt.cons(5,SortInt.cons(8,SortInt.cons(11,SortInt.nil))) : SortInt.sequence
```

Since the function insert is not mentioned in the output signature SORT, it is considered local to the body of Sort and does not appear in the structure SortInt. The body of Sort makes no reference to other functors but of course it is possible to define new functors by building on top of existing functors. For example, it would be possible to isolate the definition of sequences and functions on sequences in a functor, and then refer to this functor in the body of Sort.

The datatype declaration in SORT constrains the type sequence defined in the body of Sort to have constructors called nil and cons, and no other constructors. We can use this information outside the body of the functor to define functions over this type by case analysis, and to test values of this type for equality. The following defines a function sum of type SortInt.sequence -> int:

```
fun sum(SortInt.nil) = 0
  | sum(SortInt.cons(a,s)) = a + sum(s)
```

Although this is a very convenient notation, it relies on the fact that SortInt.sequence is defined as a datatype with a known set of constructors. It is sometimes desirable to hide such information about the representation of a type, keeping it local to the body of the functor which defines the type; this permits the representation to be changed for reasons such as time or space efficiency without changing other code which makes use of the type. The following version of SORT mentions the values cons and nil, but does not require them to be constructors:

```
signature SORT' =
   sig
      structure Elements : PO
      type sequence
      val nil : sequence
      val cons : Elements.elem * sequence -> sequence
      val sort : sequence -> sequence
   end
```

If the definition of Sort were changed to use this as output signature, then the above definition of sum would not be well-formed, even if the type sequence were defined as a datatype as above. In fact, without some additional discriminator and destructor functions (such as null, hd and tl) it would be impossible to define sum outside the body of Sort. A variation on the above would be to replace the declaration of sequence in SORT' with the line:

```
eqtype sequence
```

This exports the equality function = : sequence * sequence -> bool (which would be hidden in the case of SORT') but not the constructors.

Multi-argument functors are treated as single-argument functors in which the input signature requires a structure with multiple substructures. For example, here is a functor which takes two structures matching PO and produces another structure matching PO (the lexicographic ordering on pairs):

```
functor Lexicographic(structure X : PO
                      structure Y : PO) : PO =
   struct
      type elem = X.elem * Y.elem
      fun le((x,y),(x',y')) = if X.le(x,x')
                              then if X.le(x',x) then Y.le(y,y') else true
                              else false
   end
```

If IntPO is defined as above and BoolPO is defined as follows:

```
structure BoolPO : PO =
   struct
      type elem = bool
      fun le(true,true) = true
        | le(true,false) = false
        | le(false,b) = true
   end
```

then the functor Lexicographic may be applied to these two structures to define an order relation on (int × bool)-pairs as follows:

```
structure Lex = Lexicographic(structure X = IntPO
                              structure Y = BoolPO)
```

Then Lex.le((2,true),(2,false)) gives the value false.

When multi-argument functors are defined, it is sometimes necessary to declare that certain components of the argument structures are common to both structures. A contrived example is the following:

```
functor Wrong(structure X : PO
              structure Y : PO) : PO =
   struct
      type elem = X.elem
      fun le(a,b) = X.le(a,b) andalso Y.le(a,b)
   end
```

(andalso is logical conjunction). The definition of Wrong is ill-typed: in the definition of the function le, the variables a and b are required to be of type X.elem (because of the first conjunct) and of type Y.elem (because of the second conjunct). Some applications of Wrong (for example, to a structure in which X is IntPO and Y is like IntPO but with the opposite ordering) will be well-typed since X.elem and Y.elem are the same type, but other applications (for example, to a structure in which X is IntPO and Y is BoolPO) will be ill-typed. The input signature of the following functor includes a *sharing constraint* which restricts application to appropriate structures:

```
functor Right(structure X : PO
              structure Y : PO
              sharing type X.elem = Y.elem) : PO =
   struct
      type elem = X.elem
      fun le(a,b) = X.le(a,b) andalso Y.le(a,b)
   end
```

In this example, it was only necessary to require that X.elem and Y.elem are the same types. It is sometimes necessary to require that whole (sub)structures are the same. For example:

```
functor Strange(structure X : PO
                structure Y : PO
                sharing X = Y) : PO =
    struct
        type elem = X.elem
        fun le(a,b) = X.le(a,b) andalso Y.le(a,b)
    end
```

This functor can only be applied to structures having two identical substructures X and Y.

It is possible to use sharing constraints to make explicit the fact that parts of the argument structure of a functor are inherited by the result structure. This information can be added to the output signature of the Sort functor above as follows:

```
functor Sort'(X : PO) : sig include SORT
                            sharing Elements = X
                        end =
    struct
        structure Elements = X
        datatype sequence =
            nil
          | cons of Elements.elem * sequence
        fun insert(a,nil) = cons(a,nil)
          | insert(a,cons(b,s)) = if Elements.le(a,b) then cons(a,cons(b,s))
                                    else cons(b,insert(a,s))
        fun sort nil = nil
          | sort(cons(a,s)) = insert(a,sort s)
    end
```

The declaration **include SORT** has the same effect as repeating the declarations in the signature SORT above. The sharing constraint **sharing Elements = X** asserts that the substructure **Elements** of the result structure is identical to the argument structure.

This example exposes a subtle but important difference between the Standard ML module language and modules as used in Extended ML. In Standard ML and Extended ML, signatures serve both to impose constraints on the bodies of structures/functors and to restrict the information which is made available externally about the types and functions which are defined in structure/functor bodies. In the examples above this was used to hide local functions (such as insert in Sort) and to hide the fact that certain values are constructors (such as nil and cons in SORT'). In Standard ML, the information passed to the outside world about a structure/functor is taken to be that in its signature(s) augmented by any information about type and structure sharing which can be inferred from the body (sharing *by construction* in [MacQ 86]). Extended ML is more strict: only the information which is explicitly recorded in the signature(s) of a structure/functor is available externally. Thus, any program which is well-typed in Extended ML will be well-typed in Standard ML but not vice versa. This additional strictness is vital to allow parts of a large software system to be developed and maintained independently. The main effect of this is that it is often necessary to include explicit inheritance constraints like the one in Sort' above. Without this constraint, the information that the type Elements.elem in the structure Sort'(IntPO) is the type int would be unavailable. (This means that structures in Extended ML are actually *abstractions* in the sense of [MacQ 86], and functors are parameterized abstractions.)

# 3 The Extended ML wide-spectrum language

This section reviews the main features of the Extended ML specification/programming language. A more complete introduction to the Extended ML language appears in [ST 85]. The version of Extended ML used in this paper is different in certain details from the one presented in [ST 85] but the general motivation and ideas and the overall appearance of specifications remains the same. [SS 89] defines the syntax and some aspects of the semantics of Extended ML, and a complete formal semantics will be forthcoming.

Extended ML is intended as a vehicle for the systematic formal development of programs from specifications by means of individually-verified steps. Extended ML is called a "wide-spectrum" language since it allows all stages in the formal development process to be expressed in a single unified framework, from the initial high-level specification to the final program itself and including intermediate stages in which specification and program are intermingled. The eventual product of the formal development process is a modular program in Standard ML, and thus Standard ML (that is, the "pure" subset of Standard ML described in Section 2) is the executable sub-language of Extended ML. Earlier stages in the development of such a program are incomplete modular programs in which some parts are only specified by means of axioms rather than defined in an executable fashion by means of ML code. This allows more information to be provided in signatures (in the form of axioms specifying properties which are required to hold of any structure matching that signature), and less information to be provided in structure and functor bodies (since axioms are permitted in place of ML code).

In Section 4, a methodology is described for gradually refining such specifications to obtain programs. During the development process it is possible (and indeed normal) to use ML's module facilities to decompose a given programming task into a number of independent subtasks. This is perhaps the most novel aspect of the Extended ML methodology — its main strength lies in the support it provides for program development "in the large". Program development "in the small" is supported as well but the mechanisms provided are not very different from those of other approaches.

In the Standard ML module language, a signature acts as an interface to a program unit (structure or functor) which serves to mediate its interactions with the outside world. The signature of a structure describes the types and values which that structure makes available to the outside world. The output signature of a functor has much the same purpose, while the input signature describes what that functor requires from the outside world in order to function as required. Only those internal details of the structure/functor which are mentioned in its signature are visible to the outside world.[2] The remaining internal details may be modified at any time as long as this externally visible behaviour is maintained.

The information in a signature is sufficient for the use of Standard ML as a programming language, but when viewed as an interface specification a signature does not generally provide enough information to permit proving program correctness (for example). To make signatures more useful as interfaces of structures in program specification and development, we allow them to include axioms which put constraints on the permitted behaviour of the components of the structure. An example of such a signature is the following more informative version of the signature PO from the last section:

```
signature PO =
  sig
      type elem
      val le : elem * elem -> bool
      axiom le(x,x)
      axiom le(x,y) andalso le(y,x) => x=y
      axiom le(x,y) andalso le(y,z) => le(x,z)
  end
```

This includes the previously-unexpressible precondition which IntPO must satisfy if Sort(IntPO) is to behave as expected, namely that IntPO.le is a partial order on IntPO.elem.

Axioms are expressions of type bool. Using such an expression as an axiom amounts to an assertion that the value of the expression is true for all values of its free variables. Axioms may be built using connectives such as not, andalso, orelse and => and quantifiers such as exists and forall, and the function = may be used to compare values of any type. This is equivalent to

---

[2] As mentioned at the end of the last section, this is not quite true in Standard ML but it is true in Extended ML.

using first-order equational logic. Of course, the Standard ML code which is obtained at the end of the program development process will not contain quantifiers or use = except on types which admit equality according to Standard ML. The declaration of a type as a **datatype** amounts in logical terms to a principle of structural induction for that type, together with axioms stating that the values of two constructor terms are equal iff the terms are identical.

Formal specifications can be viewed as abstract programs. Some specifications are so completely abstract that they give no hint of an algorithm (e.g. the specification of the inverse of a matrix $A$ as that matrix $A^{-1}$ such that $A \times A^{-1} = I$) and often it is not clear if an algorithm exists at all, while other specifications are so concrete that they amount to programs (e.g. Standard ML programs, which are just equations of a certain form which happen to be executable). In order to allow different stages in the evolution of a program to be expressed in a single framework, we allow structures to contain a mixture of ML code and non-executable axioms. Functors can include axioms as well since they are simply parameterized structures. For example, a stage in the development of the functor Sort in the last section might be the following:

```
functor Sort(X : PO) : sig include SORT
                          sharing Elements = X
                       end =
   struct
       structure Elements : PO = X
       datatype sequence =
           nil
         | cons of Elements.elem * sequence
       fun append(nil,s) = s
         | append(cons(a,s1),s2) = cons(a,append(s1,s2))
       fun member(a:Elements.elem,s:sequence) = ? : bool
       axiom member(a,nil) = false
       axiom member(a,cons(a,s)) = true
       axiom a<>b => member(a,cons(b,s)) = member(a,s)
       fun insert(a:Elements.elem,s:sequence) = ? : sequence
       axiom member(a,insert(a,s))
       axiom insert(a,s) = append(s1,cons(a,s2))
                       => append(s1,s2) = s
                          andalso (member(a1,s1) => Elements.le(a1,a))
                          andalso (member(a2,s2) => Elements.le(a,a2))
       fun sort nil = nil
         | sort(cons(a,s)) = insert(a,sort s)
   end
```

In this functor declaration, the function sort has been defined in an executable fashion in terms of insert which is so far only constrained by axioms (these axioms refer to other functions which will not be required in the final version). Functions and constants which are not defined in an executable fashion are declared using the special place-holder expression ? as in the example above. This is necessary in order to declare the type of the function or constant which would normally be inferred from an executable definition by the ML system. The same construct can be used to declare a type when its representation in terms of other types has not yet been selected. It is also useful at the earliest stage in the development of a functor or structure when no body has been supplied:

```
functor Sort(X : PO) : sig include SORT
                          sharing Elements = X
                       end = ?
```

The Extended ML language is the result obtained by extending Standard ML as indicated above. That is, axioms are allowed in signatures and in structures, and the place-holder ? is allowed in place of the expression (type expression, value expression, or structure expression) on the right-hand side of declarations. Explicit signatures are required in structure declarations and explicit output signatures are required in functor declarations (in Standard ML these are optional) and the use of these signatures in typechecking is somewhat stricter than in Standard ML as discussed at the end of Section 2.2.

The examples above and those in the sequel use the notation of first-order equational logic to write axioms. This choice is rather arbitrary since the formal underpinnings of Extended ML are actually entirely independent of the choice of logic (see [ST 86] for the details; a logic suitable for use is called an *institution* [GB 84]). It is natural to choose a logic which has the Standard ML core language as a subset; this way, the development process comes to an end when all the axioms in structure and functor bodies are expressed in this executable subset.

The role of signatures as interfaces suggests that they should be regarded as descriptions of the externally observable behaviour of structures. Consider the following example:

```
signature OBJ =
    sig
        type object
    end

signature STACK =
    sig
        structure Obj : OBJ
        type stack
        val empty : stack
        val push : Obj.object * stack -> stack
        val pop : stack -> stack
        val top : stack -> Obj.object
        axiom pop(push(a,s)) = s
        axiom top(push(a,s)) = a
    end

functor Stack(O : OBJ) : sig include STACK
                             sharing Obj = O
                         end = ?

structure IntStack : STACK = Stack(struct
                                       type object = int
                                   end)
```

The purpose of the axioms in the signature STACK is to specify the behaviour of the functions defined by the functor Stack. Any implementation of these functions which satisfies the axioms in STACK will be valid. This definition of validity seems reasonable, but it turns out to be too restrictive: for instance, the usual representation of stacks using an array with a pointer to the top element of the stack will be invalid since it does not satisfy the first axiom of STACK. The reason why this representation causes no difficulties in practice is that there is no way for an external observer to detect the difference between the stacks pop(push(a,s)) and s, since equality on stacks is not provided. This implies that this axiom is not to be taken too seriously. In contrast, in IntStack (and other instantiations of Stack) it is possible to directly observe the value of top(push(a,s)) and compare it with the value of a, so the second axiom of STACK must be satisfied by any implementation of Stack. In fact, the first axiom cannot be disregarded either since it is possible in IntStack to directly observe whether or not the following equations hold (for any values of $a$ and $s$):

$$\begin{aligned} \text{top}(\text{ pop}(\text{push}(a,s))\text{ }) &= \text{top}(\text{ }s\text{ }) \\ \text{top}(\text{pop}(\text{ pop}(\text{push}(a,s))\text{ })) &= \text{top}(\text{pop}(\text{ }s\text{ })) \\ \text{top}(\text{pop}(\text{pop}(\text{ pop}(\text{push}(a,s))\text{ })))&= \text{top}(\text{pop}(\text{pop}(\text{ }s\text{ }))) \\ &\vdots \end{aligned}$$

All of these are consequences of the first axiom and so one would expect them to hold. So the first axiom is important at least insofar as it gives rise to a large number (in fact, an infinite number) of observable properties.

Because of examples like the one above, validity of implementations is defined in Extended ML in terms of satisfaction of axioms "up to behavioural equivalence" with respect to an appropriate set of "observable types". The details of this may be found in [ST 89]. The proper treatment of this issue is one of the most important facets of the design of Extended ML. However, this complication will be disregarded in this paper in the interests of simplicity of presentation; we will pretend that axioms are to be satisfied "literally", rather than only up to behavioural equivalence. Many examples, including the one in Section 5, do not require the extra generality provided by Extended ML's use of behavioural equivalence and so the language and methodology are still quite useful even when this issue is ignored.

## 4 The formal program development methodology

The starting point of formal development is a high-level requirements specification of a software system. The concept of a Standard ML functor corresponds to the informal notion of a self-contained software system. A functor may be built by composing other functors and so the scale of such a system may vary from small (like the examples in previous sections) to very large. In Extended ML, a specification of a software system is a functor with specified interfaces. The initial high-level specification will be a functor of the form:

**functor F(X : SIG) : SIG' = ?**

where SIG and SIG' are Extended ML signatures containing axioms. At later stages of development, a functor specification may include a body which is not yet composed of executable code. This is still a specification of a software system, but one in which some details of the intended implementation have been supplied.

We will not be concerned here with the difficult problem of how the initial requirements specification is obtained, or how to check that it accurately reflects the needs of the customer for whom the system is being developed. This is definitely a vital issue which needs a great deal more investigation. We assume here that a formal requirements specification in the form indicated above is provided somehow as a starting point, and ignore the step from the informal requirements of the customer to this formal specification. It is clear, however, that the formal requirements specification is the result of negotiation with the customer, and that re-negotiation will be required if it becomes necessary to change that specification in the course of the program development process.

Any non-executable Extended ML functor specification, i.e. a functor specification having a body consisting only of the placeholder ? or having a non-trivial body which is however not yet composed entirely of executable code, is regarded as a specification of a programming task. The task which is specified is (in the case of ?) to fill in a body which satisfies the functor interfaces, or (in the case of a body containing axioms) to fill in a body which satisfies the axioms in the current body.

Given a specification of a programming task, there are three ways to proceed towards a program which satisfies the specification:

**Decomposition step:** Decompose the functor into a composition of "smaller" functors, which are then regarded as separate programming tasks in their own right.

**Coding step:** Provide a functor body in the form of an abstract program containing type and value declarations and a mixture of axioms and code to define them.

**Refinement step:** Further refine an abstract program by providing a more concrete (but possibly still non-executable) version which fills in some of the decisions left open by the more abstract version.

Decomposition and coding steps are applicable to functor specifications like the one shown above in which the body consists only of the placeholder ?, while refinement steps are applicable to functor specifications which already have a body of some kind. Decomposition steps may be seen as programming (or program design) "in the large", while coding and refinement steps are programming "in the small".

Each of the three kinds of step gives rise to one or more proof obligations which can be generated mechanically from the "before" and "after" versions of the functor. Each proof obligation is a condition of the form:

$$exp_1 \cup \cdots \cup exp_n \models SIG$$

where $exp_1, \ldots, exp_n$ are Extended ML signatures or structure expressions and $SIG$ is an Extended ML signature. Discharging such a proof obligation requires showing that the axioms in the signature $SIG$ logically follow from the axioms and definitions in $exp_1, \ldots, exp_n$. A step is correct if all the proof obligations it incurs can be shown to hold. An executable Standard ML program which is obtained via a sequence of correct steps from an Extended ML specification of requirements is guaranteed to satisfy that specification. Of course, there is no need to actually do the proofs when the steps are performed; for example, they may be deferred until it is clear that a particular development path is likely to yield a satisfactory result, or until the entire development process is complete.

The details of each kind of step are given below. The example in Section 5 shows how each kind of step is used in practice to make progress during the process of developing a software system from a specification.

### 4.1 Decomposing functors

**Decomposition step** Given an Extended ML functor of the form:

    functor F(X0 : SIG0) : SIG0' = ?

we may proceed by introducing a number of additional functors:

    functor G1(X1 : SIG1) : SIG1' = ?
                    ⋮
    functor Gn(Xn : SIGn) : SIGn' = ?

and replacing the definition of F with the definition:

    functor F(X0 : SIG0) : SIG0' = strexp

where $strexp$ is a structure expression which refers to the functors G1,...,Gn. The developments of G1,...,Gn may then proceed separately.

The new definition of F is required to be a well-formed Extended ML functor definition. A number of proof obligations are incurred, one for each point in the expression $strexp$ where two modules come into contact. This includes the point where the result delivered by $strexp$ is returned as the result of F. In particular:

1. If the parameter structure X0 is used in $strexp$ in a context which demands a structure of signature $SIG$, then it is necessary to prove that $SIG0 \models SIG$.

2. If the result of an application of G$j$ is used in a context which demands a structure of signature $SIG$, then it is necessary to prove that $SIGj' \models SIG$.

3. If any other structure $STR$ (explicit structure definition or structure identifier) is used in *strexp* in a context which demands a structure of signature $SIG$, then it is necessary to prove that $STR \models SIG$. □

The best way to understand the above is to consider a simple and very typical schematic example. Let F be an Extended ML functor of the form:

```
functor F(X0 : SIG0) : SIG0' = ?
```

We may proceed by introducing two new functors:

```
functor G1(X1 : SIG1) : SIG1' = ?
functor G2(X2 : SIG2) : SIG2' = ?
```

and replacing the definition of F with the definition

```
functor F(X0 : SIG0) : SIG0' = G2(G1(X0))
```

This incurs three proof obligations:

1. Any parameter of F is a suitable parameter for G1:   $SIG0 \models SIG1$

2. Any structure delivered by G1 is a suitable parameter for G2:   $SIG1' \models SIG2$

3. Any structure delivered by G2 is a suitable result for F:   $SIG2' \models SIG0'$

Proving that $SIG0 \models SIG1$ is a matter of showing that the axioms in SIG1 logically follow from the axioms in SIG0, and likewise for the other two proof obligations.

In practice, most of the proof obligations incurred by decomposition steps are trivial to discharge by syntactic means since interfaces will almost always match exactly (i.e., in the above schematic example we will nearly always have SIG0 = SIG1, SIG1' = SIG2 and SIG2' = SIG0'). In the example in Section 5, there are four decomposition steps which give rise to a total of nineteen proof obligations. Seventeen of these are trivial because the signatures involved match syntactically, and one is trivial because there are no axioms to prove in the consequent signature. The remaining one is also trivial since all the axioms in the consequent signature appear explicitly in the antecedent signature.

## 4.2 Coding functor bodies

**Coding step** Given an Extended ML functor of the form:

```
functor F(X : SIG) : SIG' = ?
```

we may proceed by replacing the definition of F with the definition:

```
functor F(X : SIG) : SIG' = strexp
```

where *strexp* is a well-formed Extended ML functor body. This incurs a single proof obligation:

$$SIG \cup strexp \models SIG'$$

in addition to any proof obligations arising from the use of structures within *strexp*. □

## 4.3 Refining abstract code

**Refinement step** Given an Extended ML functor of the form:

    functor F(X : SIG) : SIG' = *strexp*

we may proceed by replacing the definition of F with the definition:

    functor F(X : SIG) : SIG' = *strexp'*

where *strexp'* is a well-formed Extended ML functor body. This incurs a single proof obligation:

$$\text{SIG} \cup strexp' \models strexp$$

in addition to any proof obligations arising from the use of structures within *strexp'*.   □

The above subsections have set forth three ways to proceed from a specification of a programming task towards a program which satisfies the specification, and the proof obligations which are thereby incurred. Of course, one would not expect the formal development of realistic programs to proceed in practice without backtracking, mistakes and iteration, and the Extended ML methodology does not remove the possibility of unwise design decisions. One problem is that it is often very difficult to get interface specifications right the first time. For example, when implementing a functor by decomposition into simpler functors it may well be necessary to adjust the interfaces both in order to obtain a decomposition which gives rise to "true" (i.e. provable) proof obligations and to resolve problems which arise later while implementing the simpler functors. If a decomposition has been proved correct then some changes to the interfaces may be made without affecting correctness: for example, in any of the simpler functors the output interface may be strengthened or the input interface weakened without problems. It is also possible to modify the interfaces of the functor being decomposed by weakening its output signature or strengthening its input signature. This will preserve the correctness of the decomposition, but since it changes the specification of the functor such changes must be cleared with the functor's clients (higher-level functors which use it and/or the customer). Once we have made such a change to an interface we can also change interfaces it is required to match in order to take advantage of the modification. Then, provided we are able to discharge the proof obligations referring to these interfaces, overall correctness is preserved.

The proof obligations listed above for each kind of development step are actually more strict than necessary. It is possible to loosen them by taking proper account of the ideas concerning behavioural equivalence mentioned at the end of Section 3. This allows each proof obligation above to be replaced by a condition of the form:

$$exp_1 \cup \cdots \cup exp_n \models_{OBS} SIG$$

where $\models_{OBS}$ denotes "behavioural consequence" with respect to a certain set $OBS$ of observable types. In principle, this makes the condition easier to satisfy since it only requires the observable consequences of the axioms in $SIG$ to follow from the axioms and definitions in $exp_1, \ldots, exp_n$ (see [ST 89] for full details). In practice, convenient methods for proving such conditions have not yet been established and so the proof itself is rather difficult. Since the examples at hand do not require this extra flexibility, we will use the simple but strict form of the conditions listed above.

Standard ML's module language does not permit functors to take other functors as arguments. An extension to permit this is under consideration at the present time, but some of the implications of such an extension on Extended ML have already been considered. From a methodological point of view, this extension adds considerable power; one intriguing point is that it seems to introduce a bottom-up element into Extended ML's top-down program development methodology. A more detailed discussion of this issue may be found in [SST 89].

# 5  An example

In this section the formal development process presented in the previous section is demonstrated by means of an example. Two different developments are given which begin from the same high-level Extended ML requirements specification and yield different Standard ML programs.

**Informal specification**   A symbol table in a compiler stores identifiers together with attributes of those identifiers which are determined at various stages during compilation. The following functions on symbol tables are required:

- Check whether or not an identifier is present in the symbol table.
- Add a new identifier to the symbol table and set its attributes.
- Look up the attributes of an identifier which is present. If the identifier is not present then return a default value.
- Change the attributes of an identifier which is already present. If the identifier is not present then nothing is changed.

Possible additional functions which would be useful in a compiler for a programming language with nested block structure are the following:

- Enter a new block. All of the identifiers in the symbol table are visible within the new block until replaced by local identifiers with the same name.
- Leave a block. All identifiers which were declared locally within the current block are removed from the symbol table.

These extra functions will not be considered here for the sake of simplicity, although the reader is invited to consider how their inclusion would alter the developments below.

## Step 0

The initial formal specification of the required system is given by the following Extended ML functor specification:

```
functor Symtab
    (structure X : ID
     structure Y : ATTRIB
        ) : sig include SYMTAB
                sharing Id = X and Attrib = Y
            end
    = ?
```

where ID, ATTRIB and SYMTAB are Extended ML signatures as follows:

```
signature ID =
    sig
        eqtype id
    end

signature ATTRIB =
    sig
        type attrib
        val null_attrib : attrib
    end
```

```
signature SYMTAB =
    sig
        structure Id : ID
        structure Attrib : ATTRIB
        type symtab
        val empty : symtab
        val add : Id.id * Attrib.attrib * symtab -> symtab
        val change_attrib : Id.id * Attrib.attrib * symtab -> symtab

        val present : Id.id * symtab -> bool
        axiom present(i,empty) = false
        axiom present(i,add(i',a',s)) = (i=i') orelse present(i,s)
        axiom present(i,change_attrib(i',a',s)) = present(i,s)

        val lookup : Id.id * symtab -> Attrib.attrib
        axiom lookup(i,empty) = Attrib.null_attrib
        axiom lookup(i,add(i,a,s)) = a
        axiom i<>i' => lookup(i,add(i',a',s)) = lookup(i,s)
        axiom present(i,s) => lookup(i,change_attrib(i,a,s)) = a
        axiom i<>i' => lookup(i,change_attrib(i',a',s)) = lookup(i,s)
    end
```

Our target language is the executable subset of Extended ML, namely the purely functional subset of Standard ML described in Section 2. A natural consequence of this is that the functions on symbol tables will explicitly take a symbol table as argument rather than working on a single fixed symbol table which is destructively updated by adding identifiers and changing attributes. Those functions which change the symbol table will return the modified symbol table as a result. Thus, values of type **symtab** represent states of the symbol table. The empty symbol table is represented by **empty**, and the functions **add** and **change_attrib** update the state of the symbol table by adding a new identifier (and setting its attributes) and resetting the attributes of an existing identifier, respectively. The functions **present** and **lookup** may be used for querying the current state of the symbol table. These functions check whether or not an identifier is present in the symbol table and look up the attributes of an identifier, respectively.

The parameters to the system are the type of identifiers (which is required to admit equality), the type of attributes, and a default attribute value called **null_attrib**. This means that the system will cater for any choice of these types and this value. Making the type of identifiers a parameter allows identifiers to be character strings (as usual) or something more elaborate. The informal specification does not say anything about the internal structure of attributes except that there must be some default attribute value, so it is natural to provide these as parameters to the system. The function **change_attrib** sets all the attributes of an identifier regardless of their present values; more complicated interpretations of the informal requirements are possible, but this will do for our purposes.

## Step 1

**Design decision (decomposition)** We implement **change_attrib** in terms of add. Exactly how this is done is left open for now. (Another possibility, which we will not consider, is to implement add using a function **insert** which adds a symbol without setting its attributes.)

We need two new functors:

```
functor Symtab'
   (structure X : ID
    structure Y : ATTRIB
          ) : sig include SYMTAB'
                   sharing Id = X and Attrib = Y
              end
     = ?

functor ChangeAttrib
   (X : SYMTAB'
          ) : sig include SYMTAB
                   sharing Id = X.Id and Attrib = X.Attrib
              end
     = ?
```

where SYMTAB' is exactly like SYMTAB except that the function change_attrib and the axioms which mention it are absent.

Then we can implement Symtab in terms of these functors as follows:

```
functor Symtab
   (structure X : ID
    structure Y : ATTRIB
          ) : sig include SYMTAB
                   sharing Id = X and Attrib = Y
              end
     = ChangeAttrib(Symtab'(structure X = X
                            structure Y = Y))
```

**Verification**  Typechecks okay. All interfaces match exactly so there is nothing to check.  □

## Step 2

**Design decision (coding)**  Implement the functor ChangeAttrib by coding change_attrib in terms of add in the obvious way.

```
functor ChangeAttrib
   (X : SYMTAB'
          ) : sig include SYMTAB
                   sharing Id = X.Id and Attrib = X.Attrib
              end
     = struct
           open X
           fun change_attrib(i:Id.id,a:Attrib.attrib,s:symtab) = ? : symtab
           axiom present(i,s) => change_attrib(i,a,s) = add(i,a,s)
           axiom not present(i,s) => change_attrib(i,a,s) = s
       end
```

The declaration open X includes the substructures, types and values of X in the result of ChangeAttrib. Thus, it abbreviates the following sequence of declarations:

```
structure Id : ID = X.Id
structure Attrib : ATTRIB = X.Attrib
type symtab = X.symtab
val empty = X.empty
val add = X.add
val present = X.present
val lookup = X.lookup
```

**Verification**  Typechecks okay. We have to show that

$$\text{SYMTAB}' \cup body \models \text{SYMTAB}$$

where *body* is the body of ChangeAttrib. The only non-trivial part of this involves the axioms of SYMTAB which are not in SYMTAB', namely those which mention the function change_attrib:

```
present(i,change_attrib(i',a',s)) = present(i,s)
present(i,s) => lookup(i,change_attrib(i,a,s)) = a
i<>i' => lookup(i,change_attrib(i',a',s)) = lookup(i,s)
```

The second of these follows directly from an axiom in the body of ChangeAttrib and an axiom in SYMTAB' while the first and third require simple case analyses. □

## Step 3

**Design decision (refinement)**  Convert the axioms for change_attrib into ML code. The only change required is to make the case analysis in the axioms explicit using if _ then _ else _.

```
functor ChangeAttrib
    (X : SYMTAB'
        ) : sig include SYMTAB
                sharing Id = X.Id and Attrib = X.Attrib
            end
  = struct
        open X
        fun change_attrib(i,a,s) = if present(i,s) then add(i,a,s) else s
    end
```

**Verification**  Typechecks okay. The axioms for change_attrib in the previous version of the body follow directly from the function definition in the current version of the body. □

## Pause for breath

At this point it is necessary to choose an representation of symbol tables as specified in Symtab' in terms of simpler data types. There are many possibilities, including at least the following (see [Sed 88] and similar texts for details):

1. Terms built from the constant empty using the constructor function add.

2. Sequences with identifiers kept in the order in which they are added.

3. Like (2), but with duplicates removed.

The following five choices require an additional order relation on identifiers to be supplied. Since this involves changing the original specification, it would be necessary to negotiate with the customer to see if this change is acceptable. Alternatively, if the customer is satisfied with a non-generic implementation of symbol tables in which the type of identifiers is fixed as strings of characters, the order relation need not be supplied since there is an appropriate one available.

4. Sequences with identifiers kept in ascending or descending order, with or without duplicates.

5. Like (4), but using an array in place of a sequence, with sequential search.

6. Like (5), but with binary search.

7. Ordered binary trees, with or without duplicates.

8. Balanced trees (e.g. 2–3–4 trees, AVL trees, 2–3 trees etc.).

The following two choices require an additional hash function to be supplied which takes identifiers to some given range of natural numbers. Since this involves changing the original specification, prior consultation with the customer is again required. And again, such a function need not be supplied by the customer if the type of identifiers is fixed as strings of characters.

9. Hash tables with separate chains of collisions kept in the order in which they are added, with or without duplicates.

10. Hash tables with linear probing, with rehashing into a larger table when the table becomes nearly full.

Other possibilities (again requiring modification to the original specification) are: a variation on (9) in which chains of collisions are kept in ascending or descending order; and, a variation on (10) with double hashing. Note that a variation on (10) in which the size of the table is fixed is *not* an option (assuming that the number of possible identifiers is infinite) since SYMTAB' requires symbol tables to be capable of storing an arbitarily large number of different identifiers.

In this paper we will look at just two of these possibilities: (1) and (3). The development process therefore splits at this point into two alternative development paths, which will be treated in two separate subsections.

## 5.1 Symbol tables represented as terms

The simplest way to represent symbol tables in ML is as terms built from the constant **empty** using the constructor function **add**. A similar method is applicable whenever there are no non-trivial equations inferrable between constructor terms of the type being represented. If this method is chosen then the implementation follows almost immediately from the specification of Symtab' in Step 1 above.

### Step 4

**Design decision (coding)** Implement the functor Symtab' by representing symbol tables directly as terms.

```
functor Symtab'
    (structure X : ID
     structure Y : ATTRIB
        ) : sig include SYMTAB'
                sharing Id = X and Attrib = Y
            end
    = struct
        structure Id : ID = X
        structure Attrib : ATTRIB = Y
        datatype symtab =
            empty
          | add of Id.id * Attrib.attrib * symtab
```

```
            fun present(i:Id.id,s:symtab) = ? : bool
            axiom present(i,empty) = false
            axiom present(i,add(i',a',s)) = (i=i') orelse present(i,s)

            fun lookup(i:Id.id,s:symtab) = ? : Attrib.attrib
            axiom lookup(i,empty) = Attrib.null_attrib
            axiom lookup(i,add(i,a,s)) = a
            axiom i<>i' => lookup(i,add(i',a',s)) = lookup(i,s)
        end
```

**Verification** Typechecks okay. All the axioms in SYMTAB' appear in the body of the functor, so there is nothing to prove. □

## Step 5

**Design decision (refinement)** Convert the axioms for present and lookup into ML code. No change is required to the axioms for present; the only change required to the axioms for lookup is to make the case analysis explicit using if _ then _ else _.

```
    functor Symtab'
        (structure X : ID
         structure Y : ATTRIB
          ) : sig include SYMTAB'
                  sharing Id = X and Attrib = Y
                end
    = struct
            structure Id : ID = X
            structure Attrib : ATTRIB = Y
            datatype symtab =
                empty
              | add of Id.id * Attrib.attrib * symtab

            fun present(i,empty) = false
              | present(i,add(i',a',s)) = (i=i') orelse present(i,s)

            fun lookup(i,empty) = Attrib.null_attrib
              | lookup(i,add(i',a,s)) = if i=i' then a else lookup(i,s)
        end
```

**Verification** Typechecks okay. The axioms for present and lookup in the previous version of the body follow directly from the function definitions in the current version of the body. □

All functor bodies are now expressed entirely in Standard ML, so we are finished with this development path. The functors appearing in the final program are given above under steps 1, 3 and 5. The following tree shows the dependencies between the development steps:

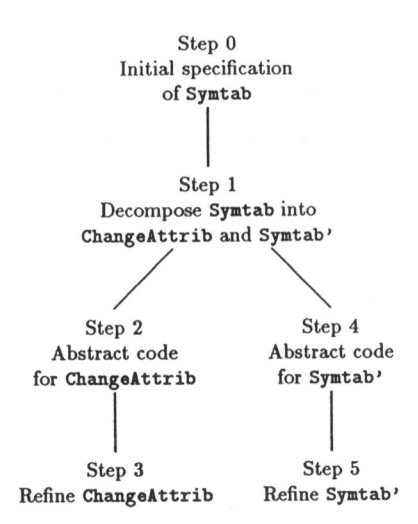

```
                Step 0
        Initial specification
              of Symtab
                  |
                Step 1
          Decompose Symtab into
         ChangeAttrib and Symtab'
               /        \
        Step 2            Step 4
     Abstract code     Abstract code
    for ChangeAttrib   for Symtab'
          |                 |
        Step 3            Step 5
    Refine ChangeAttrib  Refine Symtab'
```

## 5.2 Symbol tables represented as sequences

An alternative to the above is to represent symbol tables using sequences of (identifier × attribute)-pairs. Having selected this representation, there are several choices to be made concerning the details:

1. What dictates the order of the entries in the sequence?

    (a) Adding an entry puts it at the front of the sequence.

    (b) Adding an entry only puts it at the front of the sequence if the identifier is not already present.

    (c) The entries are kept in order of their identifiers (with respect to some order on identifiers).

2. Are duplicates removed?

    (a) Duplicates are not removed: each **add** puts an additional entry in the sequence.

    (b) Duplicates are removed.

We will choose the combination of 1(a) and 2(b) here. Recall that Steps 0–3 from the beginning of this section are still relevant to this development path.

### Step 4

**Design decision (decomposition)** We implement Symtab' in terms of sequences of (identifier × attribute)-pairs. Exactly how the functions of Symtab' are expressed in terms of the functions provided on sequences is left open for now.

We need two new functors:

```
functor SeqPairs
    (structure X : ID
     structure Y : ATTRIB
        ) : sig include SEQPAIRS
                  sharing Id = X and Attrib = Y
              end
    = ?

functor Symtab''
    (S : SEQPAIRS
        ) : sig include SYMTAB'
                  sharing Id = S.Id and Attrib = S.Attrib
              end
    = ?
```

where `SEQPAIRS` is as follows:

```
signature SEQPAIRS =
    sig
        structure Id : ID
        structure Attrib : ATTRIB
        datatype sequence =
            nil
          | cons of (Id.id * Attrib.attrib) * sequence

        val null : sequence -> bool
        axiom null nil = true
        axiom null(cons((i,a),s)) = false

        val hd : sequence -> Id.id * Attrib.attrib
        axiom hd(cons((i,a),s)) = (i,a)

        val tl : sequence -> sequence
        axiom tl(cons((i,a),s)) = s
    end
```

Then we can implement **Symtab'** in terms of these functors as follows:

```
functor Symtab'
    (structure X : ID
     structure Y : ATTRIB
        ) : sig include SYMTAB'
                  sharing Id = X and Attrib = Y
              end
    = Symtab''(SeqPairs(structure X = X
                        structure Y = Y))
```

**Verification**  Typechecks okay. All interfaces match exactly so there is nothing to check.  □

## Step 5

**Design decision (decomposition)**  We implement **Symtab''** in terms of sequences of (identifier × attribute)-pairs where no more than one pair in a sequence has the same identifier. Exactly how the functions of **Symtab''** are expressed in terms of the functions provided on such sequences is left open for now.

We need two new functors:

```
functor SeqDup
    (S : SEQPAIRS
        ) : sig include SEQDUP
                sharing Seq = S
            end
    = ?

functor Symtab'''
    (S : SEQDUP
        ) : sig include SYMTAB'
                sharing Id = S.Seq.Id and Attrib = S.Seq.Attrib
            end
    = ?
```

where SEQDUP is as follows:

```
signature SEQDUP =
    sig
        structure Seq : SEQPAIRS
        val add : (Seq.Id.id * Seq.Attrib.attrib) * Seq.sequence -> Seq.sequence
        val ismatch : Seq.Id.id * Seq.sequence -> bool
        val remove : Seq.Id.id * Seq.sequence -> Seq.sequence

        (* axioms for add *)
        axiom ismatch(i,s) => add((i,a),s) = Seq.cons((i,a),remove(i,s))
        axiom not ismatch(i,s) => add((i,a),s) = Seq.cons((i,a),s)

        (* axioms for ismatch *)
        axiom ismatch(i,Seq.nil) = false
        axiom ismatch(i,Seq.cons((i',a'),s)) = (i=i') orelse ismatch(i,s)

        (* axioms for remove *)
        axiom not ismatch(i,remove(i,s))
        local
            val member : (Seq.Id.id * Seq.Attrib.attrib) * Seq.sequence -> bool
            axiom member(e,Seq.nil) = false
            axiom member(e,Seq.cons(e',s)) = (e=e') orelse member(e,s)
        in
            axiom i<>i' => member((i',a'),remove(i,s)) = member((i',a'),s)
        end
    end
```

The axioms for **add** ensure that sequences built from **empty** and **add** contain no pairs with duplicate identifiers. The function **member** is an auxiliary function which is introduced in order to simplify the specification of **remove**. Since it is declared as local, it need not be implemented in structures which match SEQDUP.

We can now implement Symtab'' in terms of SeqDup and Symtab''' as follows:

```
functor Symtab''
    (S : SEQPAIRS
        ) : sig include SYMTAB'
                sharing Id = S.Id and Attrib = S.Attrib
            end
    = Symtab'''(SeqDup(S))
```

**Verification** Typechecks okay. All interfaces match exactly so there is nothing to check. □

## Step 6

**Design decision (coding)** Implement the functor `Symtab'''` by representing `symtab` using sequences, with `add` on symbol tables implemented by `add` (without duplicates) on sequences.

```
functor Symtab'''
    (S : SEQDUP
        ) : sig include SYMTAB'
                sharing Id = S.Seq.Id and Attrib = S.Seq.Attrib
            end
    = struct
        structure Id : ID = S.Seq.Id
        structure Attrib : ATTRIB = S.Seq.Attrib
        type symtab = S.Seq.sequence
        val empty = S.Seq.nil
        fun add(i,a,s) = S.add((i,a),s)
        val present = S.ismatch

        fun lookup(i:Id.id,s:symtab) = ? : Attrib.attrib
        axiom lookup(i,empty) = Attrib.null_attrib
        axiom lookup(i,add(i,a,s)) = a
        axiom i<>i' => lookup(i,add(i',a',s)) = lookup(i,s)
    end
```

**Verification** Typechecks okay. We have to show that

$$\text{SEQDUP} \cup body \models \text{SYMTAB}'$$

where *body* is the body of `Symtab'''`. The only non-trivial part of this involves the axioms for the function `present` in SYMTAB':

```
present(i,empty) = false
present(i,add(i',a',s)) = (i=i') orelse present(i,s)
```

The first of these follows directly from the definition of `present` in the body of `Symtab'''` and an axiom in SEQDUP. To prove the second we must first prove the following:

**Lemma** *The formula*

```
i<>i' => S.ismatch(i,S.remove(i',s)) = S.ismatch(i,s)
```

*follows from* SEQDUP ∪ *body.*

**Proof** By structural induction on the type `S.Seq.sequence`. Structural induction is valid since this type is declared (in SEQPAIRS, which is part of SEQDUP) as a datatype.    □(of Lemma)

The required result then follows by a simple case analysis.    □

## Step 7

**Design decision (refinement)** Convert the axioms for `lookup` into ML code. The only change required is to make the case analysis in the axioms explicit using `if _ then _ else _`.

```
functor Symtab'''
    (S : SEQDUP
        ) : sig include SYMTAB'
                sharing Id = S.Seq.Id and Attrib = S.Seq.Attrib
            end
  = struct
        structure Id : ID = S.Seq.Id
        structure Attrib : ATTRIB = S.Seq.Attrib
        type symtab = S.Seq.sequence
        val empty = S.Seq.nil
        fun add(i,a,s) = S.add((i,a),s)
        val present = S.ismatch
        fun lookup(i,s) = if S.Seq.null s then Attrib.null_attrib
                          else let val (i',a') = S.Seq.hd s in
                               if i=i' then a'
                               else lookup(i,S.Seq.tl s) end
    end
```

**Verification**  Typechecks okay. We have to prove that the axioms for lookup in the previous version of the body of Symtab''' follow from the function definition in the current version of the body and the axioms in SEQDUP (this contains SEQPAIRS, so the axioms there may be used as well). The relevant axioms from the previous version of Symtab''' are:

```
lookup(i,empty) = Attrib.null_attrib
lookup(i,add(i,a,s)) = a
i<>i' => lookup(i,add(i',a',s)) = lookup(i,s)
```

The first of these follows directly from the definition of lookup above and an axiom in SEQPAIRS. The other two require a simple case analysis.  □

## Step 8

**Design decision (coding)**  Implement the functor SeqDup. At this stage we convert the axioms for add and ismatch to ML code, but leave remove defined by axioms. (An alternative would be to implement SeqDup using a more general functor for duplicate-free sequences of arbitrary elements which is parameterized by the type of elements, the type of keys and the function which produces the key of an element, and use this functor for the case where elements are (identifier × attribute)-pairs, keys are identifiers, and the left projection produces the key of an element. But this would require some restructuring since we have already decided to use sequences as specified in SEQPAIRS as symbol tables.)

```
functor SeqDup
    (S : SEQPAIRS
        ) : sig include SEQDUP
                sharing Seq = S
            end
  = struct
        structure Seq : SEQPAIRS = S
        fun ismatch(i,Seq.nil) = false
          | ismatch(i,Seq.cons((i',a'),s)) = (i=i') orelse ismatch(i,s)
```

```
fun remove(i:Seq.Id.id,s:Seq.sequence) = ? : Seq.sequence
axiom not ismatch(i,remove(i,s))
local
    fun member((i:Seq.Id.id,a:Seq.Attrib.attrib),s:Seq.sequence)
                                                    = ? : bool
    axiom member(e,Seq.nil) = false
    axiom member(e,Seq.cons(e',s)) = (e=e') orelse member(e,s)
in
    axiom i<>i' => member((i',a'),remove(i,s)) = member((i',a'),s)
end

fun add((i,a),s) = if ismatch(i,s) then Seq.cons((i,a),remove(i,s))
                   else Seq.cons((i,a),s)
end
```

**Verification** Typechecks okay. The axioms for add and ismatch in SEQDUP follow directly from the function definitions in the body of SeqDup, and the axioms for remove are unchanged. □

## Step 9

**Design decision (refinement)** Supply ML code for the function remove. The local declaration of member becomes superfluous (and need not be converted to ML code) since the code for remove does not refer to member.

```
functor SeqDup
    (S : SEQPAIRS
    ) : sig include SEQDUP
            sharing Seq = S
        end
    = struct
        structure Seq : SEQPAIRS = S
        fun ismatch(i,Seq.nil) = false
          | ismatch(i,Seq.cons((i',a'),s)) = (i=i') orelse ismatch(i,s)
        fun remove(i,Seq.nil) = Seq.nil
          | remove(i,Seq.cons((i',a'),s)) = if i=i' then remove(i,s)
                                            else Seq.cons((i',a'),remove(i,s))
        fun add((i,a),s) = if ismatch(i,s) then Seq.cons((i,a),remove(i,s))
                           else Seq.cons((i,a),s)
    end
```

**Verification** Typechecks okay. We have to show that

$$\text{SEQPAIRS} \cup \text{body of SeqDup} \models \text{previous version of body of SeqDup}$$

The only axioms in the previous version of the body of SeqDup which do not appear in the present version are those for remove:

```
not ismatch(i,remove(i,s))
i<>i' => member((i',a'),remove(i,s)) = member((i',a'),s)
```

Both proofs proceed by structural induction on s and case analysis. For the second proof we are allowed to make use of the axioms for the function member in the previous version of the body of SeqDup. □

# Step 10

**Design decision (decomposition)** We implement `SeqPairs` using a more general functor `Seq` which is parameterized by the type of elements. We will use this for the case where elements are (identifier × attribute)-pairs.

We need one new functor:

```
functor Seq
    (E : ELEM
        ) : sig include SEQ
                sharing Elem = E
            end
    = ?
```

where ELEM and SEQ are as follows:

```
signature ELEM =
    sig
        type elem
    end

signature SEQ =
    sig
        structure Elem : ELEM
        datatype sequence =
            nil
          | cons of Elem.elem * sequence

        val null : sequence -> bool
        axiom null nil = true
        axiom null(cons(e,s)) = false

        val hd : sequence -> Elem.elem
        axiom hd(cons(e,s)) = e

        val tl : sequence -> sequence
        axiom tl(cons(e,s)) = s
    end
```

Then we can implement `SeqPairs` in terms of this functor as follows:

```
functor SeqPairs
    (structure X : ID
     structure Y : ATTRIB
        ) : sig include SEQPAIRS
                sharing Id = X and Attrib = Y
            end
    = struct
        structure Id : ID = X
        structure Attrib : ATTRIB = Y
        structure Elem : ELEM =
            struct
                type elem = Id.id * Attrib.attrib
            end
```

```
            structure Seq : SEQ = Seq(Elem)
            open Seq
        end
```

**Verification**  Typechecks okay. All of the structure declarations in the body of SeqPairs are trivially well-formed (in the case of Elem, this is because ELEM contains no axioms). We therefore have only to show that

$$\left( \begin{array}{l} \texttt{structure Id : ID} \\ \texttt{structure Attrib : ATTRIB} \\ \texttt{structure Elem : ELEM} \\ \texttt{structure Seq : SEQ} \\ \texttt{open Seq} \end{array} \right) \models \text{SEQPAIRS}$$

All the axioms in SEQPAIRS follow immediately.  □

## Step 11

**Design decision (coding)**  One would normally expect the functor Seq to be available in the library. In case it is not available, the axioms can be converted directly into ML code using an implementation of sequences as terms.

```
functor Seq
    (E : ELEM
    ) : sig include SEQ
            sharing Elem = E
        end
  = struct
        structure Elem : ELEM = E
        datatype sequence =
            nil
          | cons of Elem.elem * sequence
        fun null nil = true
          | null(cons(e,s)) = false
        fun hd(cons(e,s)) = e
        fun tl nil = nil
          | tl(cons(e,s)) = s
    end
```

**Verification**  Typechecks okay. All the axioms in SEQ follow immediately from the function definitions in the body of Seq. Strictly speaking, this code is invalid since the function hd is not totally defined, but any choice of value for hd nil will do (the same is true for tl nil, which we have arbitrarily given the value nil). The values of hd nil and tl nil are unimportant since the specified interface of Seq does not make any promises concerning them. However, we are operating under the global assumption that all functions are total, which means that we should ensure that some value is returned.  □

All functor bodies are now expressed entirely in Standard ML, so we are finished with this development path. The functors appearing in the final program are given above under steps 1, 3, 4, 5, 7, 9, 10 and 11. The following tree shows the dependencies between the development steps:

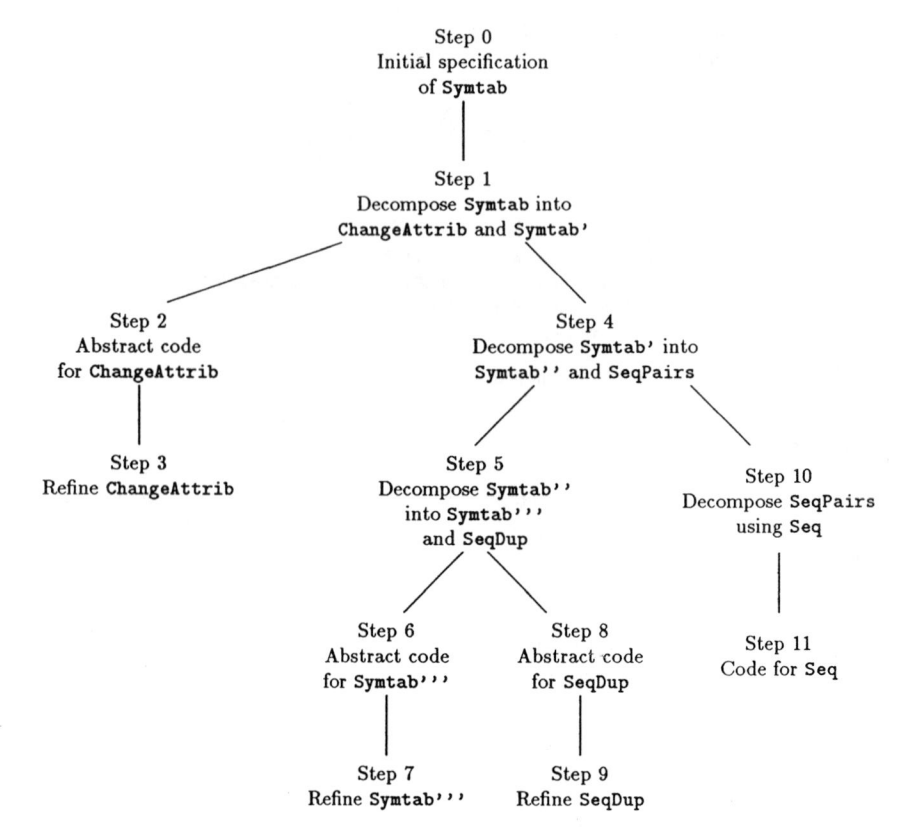

```
                    Step 0
              Initial specification
                   of Symtab
                       |
                    Step 1
              Decompose Symtab into
              ChangeAttrib and Symtab'
                  /           \
           Step 2              Step 4
        Abstract code      Decompose Symtab' into
       for ChangeAttrib    Symtab'' and SeqPairs
             |              /           \
           Step 3        Step 5          Step 10
      Refine ChangeAttrib Decompose Symtab''  Decompose SeqPairs
                        into Symtab'''         using Seq
                         and SeqDup              |
                        /        \
                    Step 6      Step 8          Step 11
                 Abstract code  Abstract code   Code for Seq
                 for Symtab'''  for SeqDup
                     |             |
                   Step 7        Step 9
               Refine Symtab'''  Refine SeqDup
```

## 6 Concluding remarks

This paper has presented the Extended ML approach to formal program development in a way which is intended to emphasize the practical aspects of formal program development while avoiding theoretical issues as much as possible. The importance of sound mathematical foundations to support the enterprise of formal program development cannot be over-emphasized, and this is one of Extended ML's strengths, but a formal program development framework should be designed in such a way that the user of the framework is not forced to be aware of these foundations.

One important feature of Extended ML which has not been stressed in this paper is the fact that the Extended ML language and methodology are practically independent of the logic used to write axioms, as well as of the form of signatures and structures (see [ST 86] for details). The notation of first-order equational logic has been used here to write axioms and signatures/structures contain types and values as in ML, but we could have used order-sorted equational logic [GM 87] and imposed a sub-type relation on types as in OBJ3 [GW 88] (although this would have been a awkward choice for producing programs in Standard ML since it is unable to cope with sub-types and coercions). The semantics of Extended ML regards executable code as a special case of axioms; e.g., Standard ML function definitions can be viewed as axioms of first-order equational logic which have the special form:

$$f(p_1) = expr_1 \wedge \cdots \wedge f(p_n) = expr_n$$

where $p_1, \ldots, p_n$ are patterns (terms containing constructors and variables only) and all the variables in $expr_j$ appear in $p_j$, for all $j \leq n$. As a practical consequence, Extended ML can be used to develop programs in other target programming languages. For example, if we switch to untyped first-order predicate logic and regard Horn clauses as the executable subset of this logic, the result is a language and methodology for developing modular Prolog programs (see [SW 87]) from specifications. Another consequence is that the present restriction to a small subset of Standard ML (excluding higher-order functions, polymorphism, references, exceptions etc.) is only necessary until a logic is developed which is able to cope with all these features adequately. Developing such a logic will not be an easy job by any means, but it is one which can be tackled separately.

The aims of Extended ML are broadly similar to those of work on rigorous program development by the VDM school (see e.g. [Jones 80]). VDM is a method for software specification and development, based on the use of explicitly-defined models of software systems, which has been widely applied in practice. However, it is *rigorous* rather than fully *formal*, and lacks formal mathematical foundations and explicit structuring mechanisms (the RAISE project [BDMP 85] is attempting to fill these gaps). In contrast, work on Extended ML builds on formal mathematical foundations with a strong emphasis on modularity and programming/design in the large; problems of practical usability are addressed, but such concerns are never allowed to take precedence over the need to maintain the soundness of the foundations. At a technical level, two advantages of the Extended ML approach (neither of which have been properly discussed here) are the use of behavioural equivalence which handles the transition between data specification and representation in a more general way than VDM's *retrieve functions*, and the independence from the underlying logical framework and target programming language mentioned above. Extended ML is primarily designed to support the development of programs from property-oriented (axiomatic) specifications rather than model-oriented specifications, but it is able to cope with model-oriented specifications as well via the use of behavioural equivalence.

Much work remains to be done. One of the most glaring omissions at present is the lack of machine-based tools to support formal program development in Extended ML. This is one of the main goals of a SERC-funded project in Edinburgh which began in May 1989. The first step will be a parser/typechecker for Extended ML specifications which will allow specifications to be checked for silly mistakes and produce abstract syntax trees in a form suitable for processing by other tools; this will be available soon. A number of theorem provers are available which are able to cope with the proofs involved in program development examples like the one in Section 5, but once one is adopted it will have to be enriched to cope with the modular structure of specifications along the lines described in [SB 83]. A component is also needed to generate proof obligations from development steps and to keep track of these and of the programming tasks which remain to be tackled. Other plans are sketched in [ST 88]. The support system will be written in Standard ML, which will allow us to experiment with the use of the techniques we advocate in developing the components of the system itself.

Acknowledgements

Much of this paper is a rehash of ideas from [ST 89]. I gratefully acknowledge the work of Andrzej Tarlecki of the Polish Academy of Sciences in our continuing collaboration on Extended ML and on topics in the foundations of algebraic specification and formal program development on which this work is based. Thanks to Edmund Kazmierczak for comments on a draft of this paper. The research reported here has been partially supported by grants from the U.K. Science and Engineering Research Council and the Polish Academy of Sciences.

# 7 References

[ Note: LNCS $n$ = Springer Lecture Notes in Computer Science, Volume $n$ ]

[BW 88] R. Bird and P. Wadler. *Introduction to Functional Programming*. Prentice-Hall (1988).

[BDMP 85] D. Bjørner, T. Denvir, E. Meiling and J. Pedersen. The RAISE project: fundamental issues and requirements. Report RAISE/DDC/EM/1/V6, Dansk Datamatic Center (1985).

[BMS 80] R. Burstall, D. MacQueen and D. Sannella. Hope: an experimental applicative language. *Proc. 1980 LISP Conference*, Stanford, California, pp. 136–143 (1980).

[GB 84] J. Goguen and R. Burstall. Introducing institutions. *Proc. Logics of Programming Workshop*, Carnegie-Mellon. LNCS 164, pp. 221–256 (1984).

[GM 87] J. Goguen and J. Meseguer. Order-sorted algebra solves the constructor-selector, multiple representation and coercion problems. *Proc. 2nd IEEE Symp. on Logic in Computer Science*, Ithaca, New York, pp. 18–29 (1987).

[GW 88] J. Goguen and T. Winkler. Introducing OBJ3. Report SRI-CSL-88-9, Computer Science Laboratory, SRI International (1988).

[Har 89] R. Harper. Introduction to Standard ML. Report ECS-LFCS-86-14, Univ. of Edinburgh. Revised edition (1989).

[HMM 86] R. Harper, D. MacQueen and R. Milner. Standard ML. Report ECS-LFCS-86-2, Univ. of Edinburgh (1986).

[HMT 89] R. Harper, R. Milner and M. Tofte. The definition of Standard ML (version 3). Report ECS-LFCS-89-81, Univ. of Edinburgh (1989).

[HW 89] P. Hudak and P. Wadler *et al*. Report on the functional programming language Haskell. Report CSC/89/R5, Univ. of Glasgow (1989).

[Jones 80] C. Jones. *Software Development: A Rigorous Approach*. Prentice-Hall (1980).

[MacQ 86] D. MacQueen. Modules for Standard ML. In: [HMM 86] (1986).

[Reade 89] C. Reade. *Elements of Functional Programming*. Addison-Wesley (1989).

[SB 83] D. Sannella and R. Burstall. Structured theories in LCF. *Proc. 8th Colloq. on Trees in Algebra and Programming*, L'Aquila, Italy. LNCS 159, pp. 377–391 (1983).

[SS 89] D. Sannella and F. da Silva. Syntax, typechecking and dynamic semantics for Extended ML. Report ECS-LFCS-89-101, Univ. of Edinburgh (1989).

[SST 89] D. Sannella, S. Sokołowski and A. Tarlecki. Toward formal development of programs from algebraic specifications: parameterisation revisited. Technical Report, Laboratory for Foundations of Computer Science, Dept. of Computer Science, Univ. of Edinburgh (to appear).

[ST 85] D. Sannella and A. Tarlecki. Program specification and development in Standard ML. *Proc. 12th ACM Symp. on Principles of Programming Languages*, New Orleans, pp. 67–77 (1985).

[ST 86] D. Sannella and A. Tarlecki. Extended ML: an institution-independent framework for formal program development. *Proc. Workshop on Category Theory and Computer Programming*, Guildford. LNCS 240, pp. 364–389 (1986).

[ST 88] D. Sannella and A. Tarlecki. Tools for formal program development: some fantasies. *LFCS Newsletter*, No. 1, pp. 10–15 (1988).

[ST 89] D. Sannella and A. Tarlecki. Toward formal development of ML programs: foundations and methodology. Report ECS-LFCS-89-71, Laboratory for Foundations of Computer Science, Dept. of Computer Science, Univ. of Edinburgh (1989); extended abstract in *Proc. Colloq. on Current Issues in Programming Languages*, Joint Conf. on Theory and Practice of Software Development (TAPSOFT), Barcelona. LNCS 352, pp. 375–389 (1989).

[SW 87] D. Sannella and L. Wallen. A calculus for the construction of modular Prolog programs. *Proc. 1987 IEEE Symp. on Logic Programming*, San Francisco, pp. 368–378 (1987). To appear in *Journal of Logic Programming*.

[Sed 88] R. Sedgewick. *Algorithms*, 2nd edition. Addison-Wesley (1988).

[Tofte 89] M. Tofte. Four lectures on Standard ML. Report ECS-LFCS-89-73, Univ. of Edinburgh (1989).

[Wik 87] Å. Wikström. *Functional Programming Using Standard ML*. Prentice-Hall (1987).

# Sheeran

The systematic design of hardware makes well the point that refinement is connected to reality at *both ends* only by general agreement. In the example of the temple, in the introduction, it is of course not possible to prove mathematically that the formal specification and Figure 2 agree; the language of elementary algebra however is expressive enough that the agreement is extremely likely. Similarly at the other end, the architect and the builders must agree on the meaning of 'width (so-and-so) and height (so-and-so)'.

At the other end of circuit design are layouts in silicon; with what mathematical expressions, in what language, do they agree? The authors here give yet more evidence that the mathematical calculus of relations is a good language for digital circuit description. Circuit components are modelled by relations between their input and output; and the 'wires' between components are just relational composition.

But description is not enough, of course. There must be a readily identifiable collection of simple relations and interconnection patterns that corresponds by general agreement to feasible circuit components and layouts. And that must lie seamlessly within a larger calculus of much more expressive relations, and operations on them, that can be used to specify the circuit behaviour in the first place. This paper draws on the authors' earlier experience with the calculus of functional programs to make that possible.

# Relations and Refinement in Circuit Design

Geraint Jones*  
Oxford University

Mary Sheeran[†]  
Glasgow University

### Abstract

A language of relations and combining forms is presented in which to describe both the behaviour of circuits and the specifications which they must meet. We illustrate a design method that starts by selecting representations for the values on which a circuit operates, and derive the circuit from these representations by a process of refinement entirely within the language.

Formal methods have always been used in circuit design. It would be unthinkable to attempt to design combinational circuits without using Boolean algebra. This means that circuit designers, unlike programmers, already use mathematical tools as a matter of course. It also means that we have a good basis on which to build higher level formal design methods. Encouraged by these observations, we have been investigating the application of formal program development techniques to circuit design.

We view circuit design as the transformation of a program describing the required behaviour into an equivalent program that is suitable for direct implementation in hardware, so there is little difference between hardware and software design. A program is implementable in hardware if it can be made from wires and circuit primitives. A program that is elegantly regular will result in a good, efficient circuit because it is communication rather than computation that is expensive in hardware. Ugly, 'spaghetti' programs are simply unacceptable as hardware designs. Because of this we have been developing notations and transformations that encourage the design of regular planar structures.

In our earlier work on the design language $\mu$FP, we modelled circuits as stream functions, and the ways that circuits are plugged together we modelled by higher-order functions [Sheer84]. These higher-order functions were chosen to have simple mathematical properties, giving a useful set of theorems for use in design by transformation. Unsurprisingly, the higher-order functions for building regular structures have attractive mathematical properties as well as being suitable for hardware implementation. This functional approach has been shown to be useful in the design of regular array architectures [Bhan89, Luk89, Jones86]. Bird and Meertens have pioneered a similar calculational approach to program design [Bird87, Bird88, Meert86, Meert89]. Their methods – which are still developing very rapidly – have been used to perform many impressive program derivations. We have presented a derivation in the Bird-Meertens style of the fast Fourier transform from its specification [Jones89].

---

*Oxford University Computing Laboratory, 11 Keble Road, Oxford OX1 3QD, England  
[†]Department of Computing Science, University of Glasgow, Glasgow G12 8QQ, Scotland

Our more recent work has been on a generalisation of the earlier functional approach. We now model circuits as binary relations between streams. The symmetry of relations gives them interesting mathematical properties [Sheer87, Luk88]. In particular, having relational inverse (converse) as a higher-order function has proved to be invaluable.

A key benefit of this approach is that it allows us to describe and reason about both combinational and sequential circuits in a uniform framework. We transform sequential circuits in much the same way as a designer uses the laws of Boolean algebra to transform combinational circuits. Reference [Sheer88] discusses the use of relations to reason about time in sequential circuits. For simplicity, we often consider only combinational circuits: circuits can then be modelled as relations between data values rather than as relations between streams of data values. Subsequent lifting to streams presents no problems and preserves the algebraic properties of the structural higher-order functions [Jones88], so the sequential design follows immediately from the combinational design.

The aim of this paper is to give a fairly gentle introduction to our calculus of relations and to show how it is used to describe and reason about data abstraction in arithmetic circuits.

## Using relations to describe hardware

We view a circuit as a binary relation between data values. In this paper we discuss only combinational circuits, so the possible data values are integers, tuples of data values and homogeneous lists of data values.

For example, the primitive relation $+$ relates a pair of integers to their integer sum. To assert that $x$ is related by the relation $R$ to $y$, we write $x\,R\,y$. Examples of true statements about $+$ are $(1,3) + 4$, $(-1,1) + 0$ and $(-3,2) + -1$. The definition of $+$ is $(x,y) + z \equiv z = x + y$. The plus sign on the right hand side is ordinary (infix) addition on the integers. We read this definition as saying that the pair $(x,y)$ is related to $x+y$ and to no other value, so $+$ is a function from its domain to its range.

The identity relation we write as $\iota$, and we name some common identities: $int$ is the identity on the integers, $nat$ on the naturals, and $bit$ on the set $\{0,1\}$. The primitive relations are

$$(x,y) + z \equiv z = x + y$$
$$(x,y) \times z \equiv z = xy$$
$$y \times 2\, z \equiv z = 2y$$
$$(x,y)\,\pi_1\, z \equiv z = x$$
$$(x,y)\,\pi_2\, z \equiv z = y$$

The only one of these relations for which there is a direct hardware implementation is the identity on bits, which corresponds to a single wire. However there are implementations for restricted versions of some relations: for example, an 'and' gate is an implementation of $\times$ restricted to bits. We assume a standard library of primitive components: in this paper, we use only logic gates, half-adders and full-adders. Primitive components usually operate on bits, so we will have to find suitable ways of representing data in terms of bits.

Circuit descriptions are built by combining relations using higher-order functions. The first two, composition and inverse, are standard operations on relations.

$$x\,(R\,;\,S)\,z \equiv \exists y.\, x\,R\,y\,\&\,y\,S\,z$$

$$\boxed{-\boxed{R}\!-\!\boxed{S}\!-}$$

Figure 1: the composition $R\,;S$

$$x\,(R^{-1})\,y \;\equiv\; y\,R\,x$$

They have the expected mathematical properties.

$$\begin{aligned}
(R\,;S)^{-1} &= S^{-1}\,;R^{-1} \\
(R^{-1})^{-1} &= R \\
R\,;(S\,;T) &= (R\,;S)\,;T
\end{aligned}$$

We draw pictures of circuits using the convention that the 'left-hand' edge is the domain and the 'right-hand' edge is the range. This means that the composition $R\,;S$ is drawn as shown in figure 1, and that taking the inverse of a circuit means flipping its picture about a vertical axis. Imagine flipping figure 1 to get a picture of $S^{-1}\,;R^{-1}$. Now you see why the equation $(R\,;S)^{-1} = S^{-1}\,;R^{-1}$ holds. Repeated composition is defined by $R^1 = R$ and $R^{n+1} = R\,;R^n$. It can be shown by induction that $(R^n)^{-1} = (R^{-1})^n$, so we feel safe in writing $R^{-n}$ for $(R^n)^{-1}$.

It is convenient to model not only circuits but types as relations. A *type* is an equivalence relation: that is a relation $R$ for which $R = R^{-1} = R^2$. If $D$ and $R$ are types for which $C = D\,;C\,;R$, we say that a relation $C$ has type $D$ to $R$, and write $C : D \sim R$. The type $D$ is a type of the domain of $C$ and the type $R$ is a type of the range of $C$. A relation may have many types.

In the composition $R\,;S$ the types of the range of $R$ and the domain of $S$ must match. If $R : A \sim B$ and $S : B \sim C$ then $R\,;S : A \sim C$. Similarly, if $R : T \sim T$ then $R^n : T \sim T$. Defining $R^0 = T$ for $R : T \sim T$ is reasonable since it means that $R^0\,;R^n = R^n\,;R^0 = R^n$.

The relation $+\,;\times 2$ is well typed and relates the pair $(x,y)$ to the value $2(x+y)$ for all integers $x$ and $y$. The relation $+\,;+$ is ill-typed. Note that

$$\begin{aligned}
\times 2\,;\times 2^{-1} &= int \\
+^{-1}\,;+ &= int
\end{aligned}$$

but that

$$\times 2^{-1}\,;\times 2 \;<\; int$$

(The ordering on relations is the subset ordering on the graphs of the relations.) Every even number is related by $\times 2^{-1}\,;\times 2$ to itself, but the odd numbers do not appear in the domain or range.

The following form, called the conjugate of $R$ by $S$, appears so often that we introduce it as a new higher-order function.

$$R \setminus S \;=\; S^{-1}\,;R\,;S$$

Taking the inverse of each side we see that

$$\begin{aligned}
(R \setminus S)^{-1} &= (S^{-1}\,;R\,;S)^{-1} \\
&= S^{-1}\,;R^{-1}\,;S \\
&= (R^{-1}) \setminus S
\end{aligned}$$

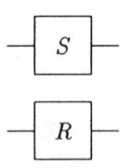

Figure 2: the parallel composition $[R, S]$

If $R : A \sim A$ and $S : A \sim C$ then $R \setminus S : C \sim A$. If in addition $Q : A \sim A$ and $S\,;S^{-1} = A$ then $(Q \setminus S)\,;(R \setminus S) = (Q\,;R) \setminus S$. We call such an $S$ a *representation* of $A$ in $C$, and $S^{-1}$ is the corresponding *abstraction*. The conjugate $R \setminus S$ implements $R$ on the concrete values in $C$, and the equality $(Q \setminus S)\,;(R \setminus S) = (Q\,;R) \setminus S$ shows that the process of implementation preserves composition.

Our next higher-order function, parallel composition (or par), makes relations on tuples. Only pairs are used in this paper but par can be extended to longer tuples.

$$(v,w)\,[R,S]\,(x,y) \;\equiv\; v\,R\,x\,\&\,w\,S\,y$$
$$[R,S]^{-1} \;=\; [R^{-1}, S^{-1}]$$
$$[R,S]\,;[T,U] \;=\; [R\,;T, S\,;U]$$
$$[R,S] \setminus [T,U] \;=\; [R \setminus T, S \setminus U]$$

Figure 2 shows how $[R, S]$ is drawn. Notice that $[int, int]$ is the identity on pairs of integers, and so is a type in its own right. For example, the relation $+$ has type $[int, int] \sim int$.

All of the examples given so far have been functions either from domain to range or from range to domain. However the parallel composition of a function and the inverse of a function is in general neither a function nor the inverse of one. For example

$$[+, +^{-1}] : [[int, int], int] \sim [int, [int, int]]$$

is neither a function from domain to range nor a function from range to domain. Because $+\,;+^{-1} > [int, int]$ and $\times 2^{-1}\,;\times 2 < int$ the relation $[+, \times 2^{-1}]\,;[+^{-1}, \times 2]$ is incomparable with $[[int, int], int]$. Beware of expecting the composition of a relation and its inverse to be an identity.

The relation $s+$ (read 'step plus') will turn out to be useful later when we consider binary arithmetic.

$$s+ \;=\; [\iota, \times 2]\,;+$$

The type of $s+$ is $[int, int] \sim int$ and each of the following is true

$$(1,0) \quad s+ \quad 1$$
$$(0,1) \quad s+ \quad 2$$
$$(1,1) \quad s+ \quad 3$$

The remaining higher-order functions that we will need make relations that operate on lists. We need some notation for describing lists of values, and some primitive relations for manipulating lists. We write $\langle\rangle$ for the empty list and $\langle x \rangle$ for the singleton list containing $x$. Appending of lists is written as $\frown$, so $\langle x \rangle \frown xs$ is the list whose first element is $x$ and

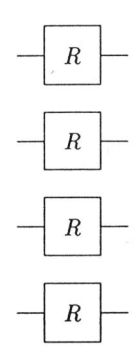

Figure 3: an instance of map $R$

whose tail is the list $xs$. All the elements of a list must be of the same type. We write just $n$ for the identity on lists of length $n$, for $n \geq 0$.

The higher-order function map builds a relation on lists from a relation on elements. If $A$ is a type then map $A$ is the type of lists of elements of type $A$ and map $\iota$ is the type of lists.

$$R : A \sim B \implies \mathsf{map}\, R : \mathsf{map}\, A \sim \mathsf{map}\, B$$
$$\langle\rangle \;\mathsf{map}\, R\; \langle\rangle$$
$$\langle x \rangle \frown xs\, (\mathsf{map}\, R)\, \langle y \rangle \frown ys \;\equiv\; x\, R\, y\; \&\; xs\, (\mathsf{map}\, R)\, ys$$

This is the relational version of the map functor that is standard in functional programming. An instance of map $R$ is shown in figure 3, and for example

$$\langle 4,4,4,4 \rangle \;\mathsf{map}\, +^{-1}\; \langle (0,4),(0,4),(0,4),(0,4) \rangle$$
$$\langle 4,4,4,4 \rangle \;\mathsf{map}\, +^{-1}\; \langle (1,3),(0,4),(0,4),(0,4) \rangle$$
$$\langle 4,4,4,4 \rangle \;\mathsf{map}\, +^{-1}\; \langle (2,2),(0,4),(0,4),(0,4) \rangle$$
$$\langle 4,4,4,4 \rangle \;\mathsf{map}\, +^{-1}\; \langle (2,2),(1,3),(0,4),(3,1) \rangle$$

and so on. The key point to note is that the $i^{th}$ elements of the lists in the domain and range are related completely independently of all the other elements of the lists. Many of the properties of map derive from its pointwise nature: it interacts nicely with composition and inverse, and hence with conjugate.

$$(\mathsf{map}\, R)^{-1} = \mathsf{map}(R^{-1})$$
$$\mathsf{map}\, R\,;\, \mathsf{map}\, S = \mathsf{map}(R\,;\, S)$$
$$\mathsf{map}\, R \setminus \mathsf{map}\, S = \mathsf{map}(R \setminus S)$$

The list constructors append, *app*, append-left, *apl*, and append-right, *apr*, are used to build lists

$$(xs, ys)\, app\, zs \;\equiv\; xs \frown ys = zs$$
$$(x, xs)\, apl\, ys \;\equiv\; \langle x \rangle \frown xs = ys$$
$$(xs, x)\, apr\, ys \;\equiv\; xs \frown \langle x \rangle = ys$$

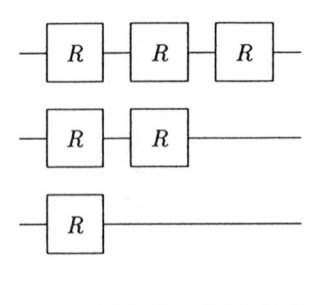

Figure 4: an instance of tri $R$

and their inverses to take lists apart. This is an example of the convenience of having available the inverse of any 'circuit'. The inverse of *app* relates a list to each pair of lists that can be concatenated to form it. The inverse of *apl* relates a non-empty list to the pair containing the head and tail of the list.

The relation *zip* interleaves two lists

$$\langle\rangle \; zip \; \langle\rangle$$
$$(\langle x\rangle \frown xs, \langle y\rangle \frown ys) \; zip \; \langle z\rangle \frown zs \;\equiv\; z = (x,y) \;\&\; (xs, ys) \; zip \; zs$$

for example

$$(\langle 1,2,3,4\rangle, \langle 5,6,7,8\rangle) \; zip \; \langle (1,5),(2,6),(3,7),(4,8)\rangle$$

Two further wiring relations are useful. They each distribute a value across a list to give a list of pairs.

$$(x, \langle\rangle) \; dstl \; \langle\rangle$$
$$(x, \langle y\rangle \frown ys) \; dstl \; \langle z\rangle \frown zs \;\equiv\; z = (x,y) \;\&\; (x, ys) \; dstl \; zs$$

and

$$(\langle\rangle, y) \; dstr \; \langle\rangle$$
$$(\langle x\rangle \frown xs, y) \; dstr \; \langle z\rangle \frown zs \;\equiv\; z = (x,y) \;\&\; (xs, y) \; dstr \; zs$$

for example

$$(1, \langle 2,3,4,5\rangle) \; dstl \; \langle (1,2),(1,3),(1,4),(1,5)\rangle$$

Triangle is a pointwise higher-order function that is slightly more complicated than map. The relation tri $R$ relates the $i^{th}$ elements of the lists in its domain and range not by $R$ but by $R^i$. Elements of lists are numbered from zero so the first elements are related by $R^0$.

$$R : T \sim T \;\Longrightarrow\; \text{tri}\, R : \text{map}\, T \sim \text{map}\, T$$
$$\langle\rangle \; \text{tri}\, R \; \langle\rangle$$
$$\langle x\rangle \frown xs \; (\text{tri}\, R) \; \langle y\rangle \frown ys \;\equiv\; x \, R^0 \, y \;\&\; xs \, (\text{map}\, R \,;\, \text{tri}\, R) \, ys$$

Relations expressed by triangle are sometimes called 'skewing' operations. The properties of tri and map depend crucially on their pointwise nature. An instance of tri $R$ is shown in figure 4.

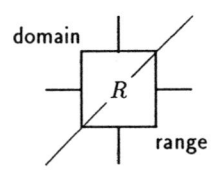

Figure 5: a picture of $R : [\iota, \iota] \sim [\iota, \iota]$

Next we introduce some higher-order functions that allow elements of lists to be combined. A convenient way to do this is to consider pair-to-pair circuits, that is circuits of type $[\iota, \iota] \sim [\iota, \iota]$, laid out according to the convention illustrated in figure 5. Two such circuits can be plugged together horizontally using the higher-order function 'beside', written $\leftrightarrow$.

$$R : [T, U] \sim [V, W] \, \& \, S : [W, X] \sim [Y, Z] \implies R \leftrightarrow S : [T, [U, X]] \sim [[V, Y], Z]$$
$$(t, (u, x)) \, R \leftrightarrow S \, ((v, y), z) \equiv \exists w. \, (t, u) \, R \, (v, w) \, \& \, (w, x) \, S \, (y, z)$$

Because $R \leftrightarrow S$ is itself pair-to-pair (or more precisely $[\iota, [\iota, \iota]] \sim [[\iota, \iota], \iota]$) circuits like $(R \leftrightarrow S) \leftrightarrow T$ make sense. Note that beside is not associative, and that $(R \leftrightarrow S) \leftrightarrow T$ is different from $R \leftrightarrow (S \leftrightarrow T)$. Perhaps the simplest example of a relation built with beside is $[\iota, \iota] \leftrightarrow [\iota, \iota]$, a relation that re-brackets three values.

$$(1, (2, 3)) \quad [\iota, \iota] \leftrightarrow [\iota, \iota] \quad ((1, 2), 3)$$
$$(1, ((2, 3), 4)) \quad ([\iota, \iota] \leftrightarrow [\iota, \iota]) \leftrightarrow [\iota, \iota] \quad (((1, 2), 3), 4)$$
$$(1, (2, (3, 4))) \quad [\iota, \iota] \leftrightarrow ([\iota, \iota] \leftrightarrow [\iota, \iota]) \quad ((1, (2, 3)), 4)$$

The higher-order function **row** uses beside repeatedly to build a linear array, an example of which is shown in figure 6. Notice that like beside, row makes a pair-to-pair relation from a smaller pair-to-pair relation.

$$R : [T, U] \sim [V, T] \implies \text{row} \, R : [T, \text{map} \, U] \sim [\text{map} \, V, T]$$
$$(t, \langle u \rangle) \, (\text{row} \, R) \, (\langle v \rangle, w) \equiv (t, u) \, R \, (v, w)$$
$$(t, \langle u \rangle \frown us) \, (\text{row} \, R) \, (\langle v \rangle \frown vs, w) \equiv (t, (u, us)) \, (R \leftrightarrow \text{row} \, R) \, ((v, vs), w)$$

For example the relation $\text{row}(+ \, ; \, +^{-1})$ relates an (integer, list of $n$ integers) pair to every (list of $n$ integers, integer) pair that has the same sum, so that

$$(1, \langle 2, 3, 4 \rangle) \, \text{row}(+ \, ; \, +^{-1}) \, (\langle 1, 2, 3 \rangle, 4)$$
$$(1, \langle 2, 3, 4 \rangle) \, \text{row}(+ \, ; \, +^{-1}) \, (\langle 1, 2, 7 \rangle, 0)$$
$$(1, \langle 2, 3, 4 \rangle) \, \text{row}(+ \, ; \, +^{-1}) \, (\langle 4, 3, 2 \rangle, 1)$$

and so on.

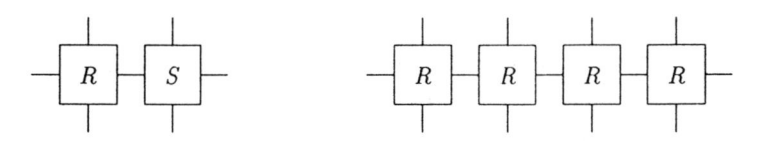

Figure 6: $R$ beside $S$ making $R \leftrightarrow S$, and an instance of row $R$

If $R : [Y, \iota] \sim [\iota, Y]$ and $D\,;A = Y$ then
$$[A, \mathsf{map}\,B]\,;\mathsf{row}\,R\,;[\mathsf{map}\,C, D] \;=\; \mathsf{row}([A, B]\,;R\,;[C, D])$$
In particular, letting $D = A = Y = \iota$ gives
$$[\iota, \mathsf{map}\,B]\,;\mathsf{row}\,R\,;[\mathsf{map}\,C, \iota] \;=\; \mathsf{row}([\iota, B]\,;R\,;[C, \iota]) \tag{1}$$
A row of $R$s can be transformed into a row of $S$s using
$$[T, U]\,;R = S\,;[V, T] \;\implies\; [T, \mathsf{map}\,U]\,;\mathsf{row}\,R = \mathsf{row}\,S\,;[\mathsf{map}\,V, T] \tag{2}$$
These theorems are proved by induction. To specify the length of a row, define
$$\mathsf{row}_n\,R = [\iota, n]\,;\mathsf{row}\,R\,;[n, \iota]$$
remembering that by $n$ we mean the identity on lists of length $n$.

Below (written $\updownarrow$) and column (written col) are useful duals of beside and row.
$$R \updownarrow S \;=\; (R^{-1} \leftrightarrow S^{-1})^{-1}$$
$$\mathsf{col}\,R \;=\; (\mathsf{row}\,R^{-1})^{-1}$$
This pattern appears often. To get the dual of a higher-order function, invert all the arguments, apply the function and invert the result. For example the dual of map is just map, but the dual of composition is reverse composition since $(F^{-1}\,;G^{-1})^{-1} = G\,;F$.

Beside, below, row and column have attractive properties.
$$(P \leftrightarrow Q) \updownarrow (R \leftrightarrow S) \;=\; (P \updownarrow R) \leftrightarrow (Q \updownarrow S)$$
$$(\mathsf{row}\,R) \updownarrow (\mathsf{row}\,S) \;=\; \mathsf{row}(R \updownarrow S)$$
and by duality
$$(\mathsf{col}\,R) \leftrightarrow (\mathsf{col}\,S) \;=\; \mathsf{col}(R \leftrightarrow S)$$
$$\mathsf{row}(\mathsf{col}\,R) \;=\; \mathsf{col}(\mathsf{row}\,R)$$

Of course not all interesting relations are pair-to-pair. It is necessary to be able to make continued versions of pair-to-value and value-to-pair relations also. The appropriate higher-order function is called reduce and is standard in functional programming. A pair-to-value relation can be viewed as a pair-to-pair relation composed with a selector. Similarly a pair-to-pair relation can be seen as a pair-to-value relation composed with an inverse selector. This allows reduce to be defined in terms of row. We concentrate on linear versions of reduce because they have attractive layouts on silicon. There are left- and right-handed versions of reduce, because there are two distinct linear ways of combining pair-to-value relations.

$$P : [X, Y] \sim X \;\implies\; \mathsf{rdl}\,P : [X, \mathsf{map}\,Y] \sim X$$
$$\mathsf{rdl}\,P \;=\; \mathsf{row}(P\,;\pi_2^{-1})\,;\pi_2$$
$$\mathsf{rdr}\,P \;=\; \mathsf{col}(P\,;\pi_1^{-1})\,;\pi_1$$

The two versions of reduce correspond to different orders of association, as shown by the following examples
$$(w, \langle x, y, z \rangle)\,(\mathsf{rdl}\,+)\,(((w + x) + y) + z)$$
$$(\langle w, x, y \rangle, z)\,(\mathsf{rdr}\,+)\,(w + (x + (y + z)))$$

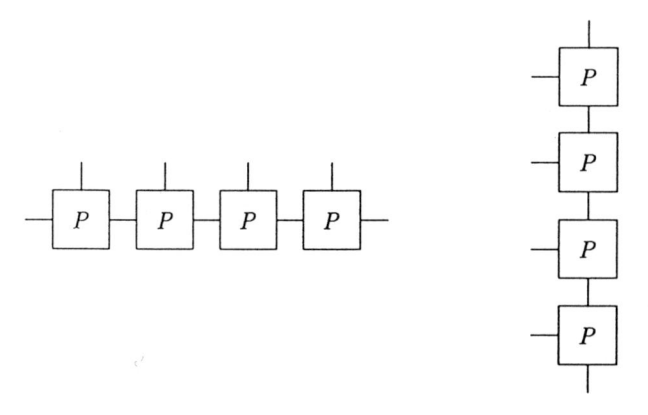

Figure 7: an instance of rdl $P$, and one of rdr $P$

which may help to explain their names. Instances of rdl $P$ and rdr $P$ are shown in figure 7. Explicitly sized versions of rdl and rdr are defined by $\text{rdl}_n P = [\iota, n]\,;\text{rdl}\, P$ and $\text{rdr}_n P = [n, \iota]\,;\text{rdr}\, P$.

Sometimes to sum the elements of a list of numbers, we will use the constant *zero* relation (defined by $x\ zero\ y \equiv x = y = 0$) to give a starting value.

$$\begin{aligned} sum &= \text{map } int\,;\pi_2^{-1}\,;[zero, \iota]\,;\text{rdl}\,+ \\ &= \text{map } int\,;\pi_1^{-1}\,;[\iota, zero]\,;\text{rdr}\,+ \end{aligned}$$

The type of *sum* is map $int \sim int$, and by the associativity of addition

$$\begin{aligned} {[int, sum]}\,;+ &= \text{rdl}\,+ \\ {[sum, int]}\,;+ &= \text{rdr}\,+ \end{aligned}$$

A version of *sum* that operates only on lists of a specific length is going to prove useful: $sum_n = n\,;sum$.

A triangle of ×2 components can be used to give weight $2^i$ to the $i^{th}$ element of the list being summed. Think of a sequence of increasing weight as a 'ladder'.

$$\begin{aligned} lad &= \text{tri} \times 2\,;sum \\ lad_n &= n\,;\text{tri} \times 2\,;sum \end{aligned}$$

For example $\langle 1, 0, 1, 1 \rangle$ *lad* 13 and $\langle 2, 3, 4, 5 \rangle$ *lad* 64.

The properties of reduce can be calculated from the properties of row. Equation 1 degenerates to

$$P : [Y, \iota] \sim Y\ \&\ D\,;A = Y \implies [A, \text{map } B]\,;(\text{rdl}\, Q)\,;D = \text{rdl}([A, B]\,;Q\,;D) \quad (3)$$

The relationship between reduce and triangle is rather more interesting.

$$[R, R]\,;S = S\,;R \implies ((\text{tri } R) \setminus apr^{-1})\,;\text{rdr}\, S = \text{rdr}([R^0, R]\,;S)$$

This could also be written using explicit sizes.

$$[R, R]\,;S = S\,;R \implies [\text{tri } R, R^n]\,;\text{rdr}_n S = \text{rdr}_n([R^0, R]\,;S)$$

For a list of length $n$ in the domain, the circuit on the left hand side of the equality needs $n(n-1)/2$ $Rs$ while the circuit on the right – although it has the same behaviour – needs only $n$ $Rs$.

A familiar example of this reduction from $O(n^2)$ to $O(n)$ is the application of Horner's rule to the evaluation of a polynomial, for example the evaluation of a binary number. Because $[\times 2, \times 2]\mathbin{;} + = +\mathbin{;} \times 2$, it follows from Horner's rule that

$$((\text{tri}\,\times 2) \setminus apr^{-1})\mathbin{;} \text{rdr}\, + \;=\; \text{rdr}([\times 2^0, \times 2]\mathbin{;} +)$$

and

$$[\text{tri}\,\times 2, \times 2^n]\mathbin{;} \text{rdr}_n\, + \;=\; \text{rdr}_n([\times 2^0, \times 2]\mathbin{;} +)$$

This last is essentially an expression of the equality

$$x_0 + 2x_1 + 2^2 x_2 + \cdots + 2^n x_n \;=\; x_0 + 2(x_1 + 2(x_2 + \cdots + 2(x_n)\cdots))$$

telling us that a binary number can be evaluated using a right reduction. In terms of the relations already introduced

$$apr\mathbin{;} lad \;=\; \text{rdr}\, s+$$

and

$$[lad_n, \times 2^n]\mathbin{;} + \;=\; \text{rdr}_n\, s+$$

Bird uses a slightly different formulation of Horner's rule to great effect in program development [Bird88].

Horner's rule and many similar skewing theorems are special cases of a very general theorem relating row and triangle. The theorem tells us how $(\text{row}\, Q) \leftrightarrow (\text{row}\, R)$ and $\text{row}(Q \leftrightarrow R)$ are related when $Q$ and $R$ satisfy certain conditions. It is discussed in reference [Jones90] where it is called the *general skewing theorem*, and will not take it any further here. An example of a consequence of the general skewing theorem is that

$$[\iota, P]\mathbin{;} Q\mathbin{;} [S, S] = [S, \iota]\mathbin{;} Q \;\Longrightarrow\; [\iota, \text{tri}\, P]\mathbin{;} \text{row}_n\, Q\mathbin{;} [\text{tri}\, S, S^n] = \text{row}_n(Q\mathbin{;} [\iota, S])$$

This can be used to prove that

$$\begin{aligned}[\iota, \text{tri}\,\times 2]\mathbin{;} \text{row}_n(+\mathbin{;} +^{-1})\mathbin{;} [\text{tri}\,\times 2^{-1}, (\times 2^n)^{-1}] &\;=\; \text{row}_n(+\mathbin{;} +^{-1}\mathbin{;} [\iota, \times 2^{-1}]) \\ &\;=\; \text{row}_n(+\mathbin{;} s+^{-1}) \end{aligned} \qquad (4)$$

This will prove useful in the design of arithmetic circuits. Another consequence of the general skewing theorem – this time without any triangles at all – tells us that under certain conditions a reduce-left and the inverse of a reduce-right can be combined to make a row. For example, it can be shown that $\text{rdl}_n\, +\mathbin{;} (\text{rdr}_n\, +)^{-1} = \text{row}_n(+\mathbin{;} +^{-1})$. Since $\text{rdl}_n\, + = [\iota, sum_n]\mathbin{;} +$ and $\text{rdr}_n\, + = [sum_n, \iota]\mathbin{;} +$, this can be rewritten as

$$([\iota, sum_n]\mathbin{;} +)\mathbin{;} ([sum_n, \iota]\mathbin{;} +)^{-1} \;=\; \text{row}_n(+\mathbin{;} +^{-1}) \qquad (5)$$

## Abstraction and representation

An abstraction relates a concrete value to the abstract value that it represents. Recall that we say that $abs : C \sim A$ is an abstraction from concrete type $C$ onto the abstract type $A$ provided that

$$abs^{-1}\mathbin{;} abs \;=\; A$$

The converse $abs^{-1} : A \sim C$ is the corresponding representation relation. All of our higher-order functions except inverse and conjugate preserve this property. So for example if $a$ is an abstraction of type $[S,T] \sim T$, then $\mathsf{rdl}\,a$ is an abstraction of type $[S, \mathsf{map}\,T] \sim T$. In dealing with arithmetic circuits, we will be considering abstractions onto the integers and the naturals.

The relations $+$ and $s+$ are abstractions of type $[int, int] \sim int$. An example of a relation that might look like an abstraction onto the integers but is not is $[bit, \times 2\,;\, \times 2]\,;\,+$. Not all integers are in its range; for instance, the integer 3 is not represented; however the very similar $[int, \times 2\,;\, \times 2]\,;\,+$ is an abstraction onto the integers.

The list constructors, $apl$ and $apr$ and their inverses are all abstractions, for $apl^{-1}\,;\, apl$ and $apr^{-1}\,;\, apr$ are both the type of non-empty lists, $apl\,;\, apl^{-1}$ is the type of (value, list) pairs and $apr\,;\, apr^{-1}$ is the type of (list, value) pairs. The relation $sum$ is an abstraction from $\mathsf{map}\,int$ to $int$. The relations $\mathsf{rdl}+\,:\,[int, \mathsf{map}\,int] \sim int$ and $\mathsf{rdr}+\,:\,[\mathsf{map}\,int, int] \sim int$ are both abstractions onto the integers. Because $\times 2$ is an abstraction onto the *even* integers, not the integers, $\mathsf{tri}\,\times 2$ is an abstraction onto lists of even integers. However $lad = \mathsf{tri}\,\times 2\,;\,sum$ is an abstraction onto the integers, as are $[int, lad]\,;\,+$, and $[int, lad]\,;\,s+$ and $[lad_n, int\,;\, \times 2^n]\,;\,+$.

Restricted versions of integer abstractions are used to make abstractions onto the naturals. For example, $bin$ relates a binary number (a list of bits, least significant bit first) to the natural number that it represents.

$$bin = \mathsf{map}\,bit\,;\,lad$$
$$bin_n = \mathsf{map}\,bit\,;\,lad_n$$

The abstraction for a (bit-carry, $n$-bit binary) pair is just $[bit, bin_n]\,;\,+$ and that for an ($n$-bit binary, carry-out bit) pair is $[bin_n, bit\,;\, \times 2^n]\,;\,+$. A carry-save number is like a binary number in which each bit is replaced by a pair of equal-weight bits, described by the abstractions

$$csv = \mathsf{map}([bit, bit]\,;\,+)\,;\,lad$$
$$csv_n = \mathsf{map}([bit, bit]\,;\,+)\,;\,lad_n$$

One can also represent a number as a ladder of bit-steps.

$$scsv = \mathsf{map}([bit, bit]\,;\,s+)\,;\,lad$$
$$scsv_n = \mathsf{map}([bit, bit]\,;\,s+)\,;\,lad_n$$

Because $+\,;\,\times 2 = [\times 2, \times 2]\,;\,+$ and $zip\,;\,\mathsf{map}+\,;\,sum_n = [sum_n, sum_n]\,;\,+$, two ladders can be zipped together

$$[lad_n, lad_n]\,;\,+ \;=\; zip\,;\,\mathsf{map}+\,;\,lad_n$$

This means that one way to add two binary numbers is to interleave them and regard the result as a carry-save number

$$\begin{aligned}
[bin_n, bin_n]\,;\,+ &= [\mathsf{map}\,bit\,;\,lad_n, \mathsf{map}\,bit\,;\,lad_n]\,;\,+ \\
&= [\mathsf{map}\,bit, \mathsf{map}\,bit]\,;\,zip\,;\,\mathsf{map}+\,;\,lad_n \\
&= zip\,;\,\mathsf{map}[bit, bit]\,;\,\mathsf{map}+\,;\,lad_n \\
&= zip\,;\,csv_n
\end{aligned} \qquad (6)$$

Similarly

$$[bin_n, bin_n] ; s+ = zip ; scsv_n \qquad (7)$$

and

$$\begin{aligned}[bin_n, csv_n] ; + &= [\text{map } bit ; lad_n, \text{map}([bit, bit] ; +) ; lad_n] ; + \\ &= [\text{map } bit, \text{map}([bit, bit] ; +)] ; zip ; \text{map} + ; lad_n \\ &= zip ; \text{map}([bit, [bit, bit] ; +] ; +) ; lad_n \end{aligned} \qquad (8)$$

Notice that × is just as good an abstraction as +. The product of two binary numbers can be expressed in terms of some wiring and binary-bit products.

$$[bin_n, bin_n] ; \times = dstl ; \text{map}([bin_n, bit] ; \times) ; lad_n \qquad (9)$$

The product of a binary number and a bit can be expressed in terms of wiring and bit-bit products.

$$\begin{aligned}[bin_n, bit] ; \times &= dstr ; \text{map}([bit, bit] ; \times) ; lad_n \\ &= dstr ; \text{map } andgate ; bin_n \\ &\text{where } andgate = [bit, bit] ; \times ; bit \end{aligned} \qquad (10)$$

Note that the product of two bits is implemented by an 'and' gate, written *andgate*.

## Designing circuits to change representations

Many signal processing circuits can be viewed as changers of representation. If $p : P \sim A$ and $q : Q \sim A$ are abstractions onto the same type $A$, then $p^{-1} ; (p ; q^{-1}) ; q = A$ and $p ; q^{-1} : P \sim Q$ relates equivalent (concrete) values from $P$ and $Q$, that is representatives of abstract values (of the type $A$) which are equivalent under the relation $A$.

This makes it easy to write down the original specification of a circuit as a changer of representation – the composition of one abstraction with the inverse of another. Theorems about the higher-order functions are then used to massage the description into something that is more easily implementable. The final circuit must use components that can be built in hardware. Such components generally perform small scale changes of representation, so the aim is to work out how to implement a large scale change of representation in terms of smaller ones.

For example, the addition of a carry-in to an weighted $n$-list to give a weighted $n$-list and carry-out is a change from the representation described by $[\iota, lad_n] ; +$ to that described by $[lad_n, \times 2^n] ; +$, so it is

$$\begin{aligned}&([\iota, lad_n] ; +) ; ([lad_n, \times 2^n] ; +)^{-1} \\ &= \{ \text{ def. } lad_n \} \\ &\quad [\iota, \text{tri} \times 2] ; [\iota, sum_n] ; + ; +^{-1} ; [sum_n^{-1}, \iota] ; [\text{tri} \times 2^{-1}, (\times 2^n)^{-1}] \\ &= \{ \text{ equation 5 } \} \\ &\quad [\iota, \text{tri} \times 2] ; \text{row}_n(+, +^{-1}) ; [\text{tri} \times 2^{-1}, (\times 2^n)^{-1}] \\ &= \{ \text{ equation 4 } \} \\ &\quad \text{row}_n(+ ; s+^{-1}) \end{aligned} \qquad (11)$$

This equation expresses a large scale change of representation as a row of smaller ones. Composing $[\iota, \mathsf{map} \times 2]$ on the left of each side of the equality gives

$$([\iota, \mathsf{map} \times 2 \,;\, lad_n] \,;\, +) \,;\, ([lad_n, \times 2^n] \,;\, +)^{-1} = \{\text{ equation 11 }\}$$
$$[\iota, \mathsf{map} \times 2] \,;\, \mathsf{row}_n(+ \,;\, s+^{-1})$$
$$= \{\text{ equation 2 }\}$$
$$\mathsf{row}_n([\iota, \times 2] \,;\, + \,;\, s+^{-1})$$
$$= \{\text{ def. } s+ \}$$
$$\mathsf{row}_n(s+ \,;\, s+^{-1})$$

and since $\mathsf{map} \times 2 \,;\, lad_n = lad_n \,;\, \times 2$ and $[\iota, \times 2] \,;\, + = s+$

$$([\iota, lad_n] \,;\, s+) \,;\, ([lad_n, \times 2^n] \,;\, +)^{-1} = \mathsf{row}_n(s+ \,;\, s+^{-1}) \qquad (12)$$

Taking inverses on both sides of equations 11 and 12 gives

$$([lad_n, \times 2^n] \,;\, +) \,;\, ([\iota, lad_n] \,;\, +)^{-1} = \mathsf{col}_n(s+ \,;\, +^{-1}) \qquad (13)$$

and

$$([lad_n, \times 2^n] \,;\, +) \,;\, ([\iota, lad_n] \,;\, s+)^{-1} = \mathsf{col}_n(s+ \,;\, s+^{-1}) \qquad (14)$$

Equations 11 to 14 represent patterns of computation that appear again and again in arithmetic circuits. To design a circuit that manipulates natural numbers or bits, we specialise by restricting the domain and range of one of these general representation changers.

## Examples: adders

The addition of a carry-in bit to a binary number, giving a binary number and carry-out, matches the pattern of equation 11.

$$([bit, bin_n] \,;\, +) \,;\, ([bin_n, bit \,;\, \times 2^n] \,;\, +)^{-1}$$
$$= \{\text{ def. } bin, \text{ and that } bit \text{ is a type }\}$$
$$[bit, \mathsf{map}\, bit] \,;\, [\iota, lad_n] \,;\, + \,;\, ([lad_n, \times 2^n] \,;\, +)^{-1} \,;\, [\mathsf{map}\, bit, bit]$$
$$= \{\text{ equation 11 }\}$$
$$[bit, \mathsf{map}\, bit] \,;\, \mathsf{row}_n(+ \,;\, s+^{-1}) \,;\, [\mathsf{map}\, bit, bit] \qquad (15)$$

The next step is to push the constraints on the values in the domain and range of the circuit inside the row. If we know that the circuit is $\mathsf{row}(C)$ for some $C$, we can proceed to find an implementation of $C$. The approach is hierarchical in that we can always choose $C$ to be itself a change of representation. The decomposition continues until we reach components that can be implemented directly in hardware.

$$[bit, \mathsf{map}\, bit] \,;\, \mathsf{row}_n(+ \,;\, s+^{-1}) \,;\, [\mathsf{map}\, bit, bit]$$
$$= \{\text{ equation 1 }\}$$
$$[bit, \mathsf{map}\, \iota] \,;\, \mathsf{row}_n([\iota, bit] \,;\, + \,;\, s+^{-1} \,;\, [bit, \iota]) \,;\, [\mathsf{map}\, \iota, bit]$$
$$= \{\text{ equation 2 and } [bit, bit] \,;\, + \,;\, s+^{-1} \,;\, [bit, \iota] = [bit, bit] \,;\, + \,;\, s+^{-1} \,;\, [bit, bit] \}$$
$$\mathsf{row}_n([bit, bit] \,;\, + \,;\, s+^{-1} \,;\, [bit, bit]) \qquad (16)$$

The justification for the last step just says that if both values in the domain and the first value of the range of $(+\ ;s+^{-1})$ are bits, then the second element of the range must also be a bit. We can stop here because the component of the row, $[bit, bit]\ ;+\ ;s+^{-1}\ ;[bit, bit]$, is just a half-adder. It relates a pair of equal weight bits to a bit-step (the sum and carry) representing the same value drawn from $\{0, 1, 2\}$.

We conclude from equations 15 and 16 that

$$([bit, bin_n]\ ;+)\ ;([bin_n, bit\ ;\times 2^n]\ ;+)^{-1}$$
$$=\ \mathsf{row}_n\ \mathit{halfadd}$$
$$\text{where } \mathit{halfadd} = [bit, bit]\ ;+\ ;s+^{-1}\ ;[bit, bit]$$

To add a bit to a carry-save number yielding a binary number and carry-out bit, we again constrain the pattern described in equation 11. The same sequence of laws is used to push the constraints inside the row.

$$([bit, csv_n]\ ;+)\ ;([bin_n, bit\ ;\times 2^n]\ ;+)^{-1}$$
$$=\{\text{def. } bin,\ csv\,\}$$
$$\quad [bit, \mathsf{map}([bit, bit]\ ;+)\ ;lad_n]\ ;+\ ;+^{-1}\ ;[lad_n^{-1}\ ;\mathsf{map}\,bit, (\times 2^n)^{-1}\ ;bit]$$
$$=\{\text{equation 11}\,\}$$
$$\quad [bit, \mathsf{map}([bit, bit]\ ;+)]\ ;\mathsf{row}_n(+\ ;s+^{-1})\ ;[\mathsf{map}\,bit, bit]$$
$$=\{\text{equation 1}\,\}$$
$$\quad [bit, \mathsf{map}\,\iota]\ ;\mathsf{row}_n([\iota, [bit, bit]\ ;+]\ ;+\ ;s+^{-1}\ ;[bit, \iota])\ ;[\mathsf{map}\,\iota, bit]$$
$$=\{\text{equation 2, types}\,\}$$
$$\quad \mathsf{row}_n([bit, [bit, bit]\ ;+]\ ;+\ ;s+^{-1}\ ;[bit, bit])$$

We can stop here because the component of the row, $[bit, [bit, bit]\ ;+]\ ;+\ ;s+^{-1}\ ;[bit, bit]$, is a full-adder. It relates an equal weight pair of a bit and a pair of equal weight bits to a bit-step (sum and carry) representing the same element of $\{0, 1, 2, 3\}$. The conclusion is that

$$([bit, csv_n]\ ;+)\ ;([bin_n, bit\ ;\times 2^n]\ ;+)^{-1}$$
$$=\ \mathsf{row}_n\ \mathit{fulladd} \tag{17}$$
$$\text{where } \mathit{fulladd} = [bit, [bit, bit]\ ;+]\ ;+\ ;s+^{-1}\ ;[bit, bit]$$

The circuit that adds a bit and two binary numbers to give a binary number and a carry-out bit also uses a row of full-adders.

$$([bit, [bin_n, bin_n]\ ;+]\ ;+)\ ;([bin_n, bit\ ;\times 2^n]\ ;+)^{-1}$$
$$=\{\text{equation 6}\,\}$$
$$\quad [bit, zip\ ;csv_n]\ ;+\ ;([bin_n, bit\ ;\times 2^n]\ ;+)^{-1}$$
$$=\{\text{equation 17}\,\}$$
$$\quad [\iota, zip]\ ;\mathsf{row}_n\ \mathit{fulladd}$$

Both $\mathit{halfadd}$ and $\mathit{fulladd}$ are functions from domain to range. If $+\ ;s+^{-1}$ is to be restricted to be a function from domain to range then the first component in the range must be restricted to be a bit, as indeed it is in both the full-adder and the half-adder.

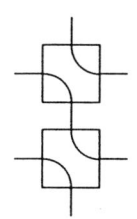

Figure 8: $[bit, bit] \updownarrow [bit, bit]$

This means that if a relation of the form $\text{row}_n(+\,;\,s+^{-1})$ is to be a function from domain to range, then the first component in its range must be a binary number.

Full-adders are also useful when adding binary and carry-save numbers.

$[bin_n, csv_n]\,;\,+$
$\quad = \{\text{ equation 8 }\}$
$\qquad zip\,;\,\text{map}([bit,[bit,bit]\,;\,+]\,;\,+)\,;\,lad_n$
$\quad = \{\,([bit,[bit,bit]\,;\,+]\,;\,+) : [bit,[bit,bit]] \sim (s+^{-1}\,;\,[bit,bit]\,;\,s+)\,\}$
$\qquad zip\,;\,\text{map}([bit,[bit,bit]\,;\,+]\,;\,+\,;\,s+^{-1}\,;\,[bit,bit]\,;\,s+)\,;\,lad_n$
$\quad = \{\text{ def. } fulladd \text{ and } scsv_n\,\}$
$\qquad zip\,;\,\text{map}\,fulladd\,;\,scsv_n \qquad\qquad\qquad\qquad\qquad\qquad (18)$

So one can add a binary number to a carry-save using just wiring and full-adders provided one regards the output – in this case the range – as a ladder of bit-steps.

Sometimes wiring alone can implement a change of representation, for example

$$[bit, \iota] \;=\; [bit, \iota]\,;\,s+\,;\,s+^{-1}\,;\,[bit, \iota] \qquad (19)$$

Similarly using $bsh\,;\,[\iota,+]\,;\,s+ = [s+,\iota]\,;\,s+$, where $bsh = [bit,bit] \updownarrow [bit,bit]$ is the wiring relation illustrated in figure 8, it can be shown that

$$[scsv_n, bit\,;\,\times 2^n]\,;\,+ \;=\; \text{col}_n\,bsh\,;\,[bit, csv_n]\,;\,s+$$

Now composing $s+^{-1}\,;\,[bit, csv_n^{-1}]$ on the right of each side and using equation 19 yields

$$[scsv_n, bit\,;\,\times 2^n]\,;\,+\,;\,s+^{-1}\,;\,[bit, csv_n^{-1}] \;=\; \text{col}_n\,bsh\,;\,[bit, csv_n\,;\,csv_n^{-1}] \qquad (20)$$

# Example: a multiplier

Finally to demonstrate that circuits other than adders fit into this framework, we outline the design a binary multiplier. It turns out to be easier to design a multiplier with a carry-in and carry-out in some suitable representation. For simplicity we design an $n$-bit by $n$-bit multiplier, although similar techniques would be used in an $n$-bit by $m$-bit multiplier. The initial specification is

$$([abs, [bin_n, bin_n]\,;\,\times]\,;\,+)\,;\,([bin_n, abs\,;\,\times 2^n]\,;\,+)^{-1}$$

where the abstraction *abs* has yet to be chosen. It is assumed that *abs* relates a fixed-length value to the naturals. The maximum carry-in or carry-out that needs to be represented is calculated by solving for $c$ in

$$c + (2^n - 1)(2^n - 1) = (2^n - 1) + 2^n c$$

The solution is $c = 2^n - 2$, so, possible choices for *abs* are $bin_n$ and $csv_{n-1}$, which can represent 0 to $2^n - 2$. Choosing $bin_n$ gives the multiplier described in reference [Sheer85]. Here we will choose $csv_{n-1}$ and design a multiplier similar to that presented in reference [McCan82]. Assuming $n > 0$, the specification is now

$$[csv_{n-1}, [bin_n, bin_n] ; \times] ; + ; +^{-1} ; [bin_n^{-1}, (\times 2^n)^{-1} ; csv_{n-1}^{-1}]$$

The first two steps are the same as in all previous examples. The specification is massaged into a form that matches one of the standard patterns represented by equations 11 to 14 and then the constraints on the domain and range are pushed inside the row or column that results.

$[csv_{n-1}, [bin_n, bin_n] ; \times] ; + ; +^{-1} ; [bin_n^{-1}, (\times 2^n)^{-1} ; csv_{n-1}^{-1}]$
$= \{$ equation 9 and def. *bin* $\}$
    $[csv_{n-1}, dstl ; \mathsf{map}([bin_n, bit] ; \times) ; lad_n] ; + ;$
    $+^{-1} ; [lad_n^{-1} ; \mathsf{map}\, bit, (\times 2^n)^{-1} ; csv_{n-1}^{-1}]$
$= \{$ equation 11 $\}$
    $[csv_{n-1}, dstl ; \mathsf{map}([bin_n, bit] ; \times)] ; \mathsf{row}_n(+ ; s+^{-1}) ; [\mathsf{map}\, bit, csv_{n-1}^{-1}]$
$= \{$ equation 10 $\}$
    $[\iota, dstl ; \mathsf{map}(dstr ; \mathsf{map}\, andgate)] ;$
    $[csv_{n-1}, \mathsf{map}\, bin_n] ; \mathsf{row}_n(+ ; s+^{-1}) ; [\mathsf{map}\, bit, csv_{n-1}^{-1}]$
$= \{$ equation 1 $\}$
    $[\iota, dstl ; \mathsf{map}(dstr ; \mathsf{map}\, andgate)] ;$
    $[csv_{n-1}, \iota] ; \mathsf{row}_n([\iota, bin_n] ; + ; s+^{-1} ; [bit, \iota]) ; [\iota, csv_{n-1}^{-1}]$

Considering the sizes of the numbers that are related by $[csv_{n-1}, bin_n] ; + ; s+^{-1} ; [bit, \iota]$ shows that any number in the second component of the range of this relation must be no bigger than $(2(2^{n-1} - 1) + (2^n - 1))/2 = 2(2^{n-1} - 1)$ and so has an $(n-1)$-bit carry-save representation. This means that

$[csv_{n-1}, bin_n] ; + ; s+^{-1} ; [bit, \iota]$
$= \ [csv_{n-1}, bin_n] ; + ; s+^{-1} ; [bit, \iota] ; [\iota, csv_{n-1}^{-1} ; csv_{n-1}]$
$= \ [csv_{n-1}, bin_n] ; + ; s+^{-1} ; [bit, csv_{n-1}^{-1}] ; [\iota, csv_{n-1}] \qquad (21)$

so we conclude that

$([csv_{n-1}, [bin_n, bin_n] ; \times] ; +) ; ([bin_n, csv_{n-1} ; \times 2^n] ; +)^{-1}$
$= \{$ equations 2 and 21 $\}$
    $[\iota, dstl ; \mathsf{map}(dstr ; \mathsf{map}\, andgate)] ;$
    $\mathsf{row}_n([csv_{n-1}, bin_n] ; + ; s+^{-1} ; [bit, csv_{n-1}^{-1}]) ;$
    $[\iota, csv_{n-1} ; csv_{n-1}^{-1}]$
$= \ [\iota, dstl ; \mathsf{map}(dstr ; \mathsf{map}\, andgate)] ; \mathsf{row}_n\, cell ; [\iota, csv_{n-1} ; csv_{n-1}^{-1}] \qquad (22)$
    where $cell = [csv_{n-1}, bin_n] ; + ; s+^{-1} ; [bit, csv_{n-1}^{-1}]$

and the next step is to find an implementation for *cell*. We have already solved a very similar problem. The addition of an $(n-1)$-bit carry-save number to an $n$-bit binary number can be recast as the addition of an $(n-1)$-bit carry-save number, an $(n-1)$-bit binary number and the most significant bit of the $n$-bit binary number.

$$[csv_{n-1}, bin_n]\,;\,+ \\ = w\,;\,[[bin_{n-1}, csv_{n-1}]\,;\,+, bit\,;\,\times 2^{n-1}]\,;\,+ \qquad (23)$$
$$\text{where } (xs, ys \frown \langle y \rangle)\,w\,((as, bs), c) \equiv as = ys\ \&\ bs = xs\ \&\ c = y$$

So *cell* can be rewritten into a form that is more suitable for implementation in hardware.

$$\begin{aligned}
cell &= \{\text{equation 23}\} \\
&\quad w\,;\,[[bin_{n-1}, csv_{n-1}]\,;\,+, bit\,;\,\times 2^{n-1}]\,;\,+\,;\,s+^{-1}\,;\,[bit, csv_{n-1}^{-1}] \\
&= \{\text{equation 18}\} \\
&\quad w\,;\,[zip\,;\,\text{map}\,fulladd, \iota]\,;\,[scsv_{n-1}, bit\,;\,\times 2^{n-1}]\,;\,+\,;\,s+^{-1}\,;\,[bit, csv_{n-1}^{-1}] \\
&= \{\text{equation 20}\} \\
&\quad w\,;\,[zip\,;\,\text{map}\,fulladd, \iota]\,;\,\text{col}_{n-1}\,bsh\,;\,[\iota, csv_{n-1}\,;\,csv_{n-1}^{-1}]
\end{aligned}$$

The only part of this design that is not directly implementable is the composition $csv_{n-1}\,;\,csv_{n-1}^{-1}$, which is the (concrete) type of $(n-1)$-bit carry-save representations. It is the relation that identifies any pair of $(n-1)$-bit carry-save representations of the same number – do not be misled into thinking that it is an identity! The best way to deal with it will be to push it out to the edge of the circuit. Because $csv_{n-1}$ is an abstraction, $csv_{n-1}\,;\,csv_{n-1}^{-1}\,;\,csv_{n-1} = csv_{n-1}$ and so $[csv_{n-1}\,;\,csv_{n-1}^{-1}, \iota]\,;\,cell = cell$, just by inspection of the definition of *cell*. It follows that

$$[csv_{n-1}\,;\,csv_{n-1}^{-1}, \iota]\,;\,cell\ =\ cell'\,;\,[\iota, csv_{n-1}\,;\,csv_{n-1}^{-1}]$$
$$\text{where } cell' = w\,;\,[zip\,;\,\text{map}\,fulladd, \iota]\,;\,\text{col}_{n-1}\,bsh$$

so by equation 2

$$[csv_{n-1}\,;\,csv_{n-1}^{-1}, \iota]\,;\,\text{row}_{n-1}\,cell\ =\ \text{row}_{n-1}\,cell'\,;\,[\iota, csv_{n-1}\,;\,csv_{n-1}^{-1}]$$

and substituting this back into equation 22

$$\begin{aligned}
([csv_{n-1}, [bin_n, bin_n]\,;\,\times]\,;\,+)\,;\,&([bin_n, csv_{n-1}\,;\,\times 2^n]\,;\,+)^{-1} \\
&= [\iota, dstl\,;\,\text{map}(dstr\,;\,\text{map}\,andgate)]\,;\,\text{row}_n\,cell'\,;\,[\iota, (csv_{n-1}\,;\,csv_{n-1}^{-1})^2] \\
&= [\iota, dstl\,;\,\text{map}(dstr\,;\,\text{map}\,andgate)]\,;\,\text{row}_n\,cell'\,;\,[\iota, csv_{n-1}\,;\,csv_{n-1}^{-1}]
\end{aligned}$$

Again everything except the $csv_{n-1}\,;\,csv_{n-1}^{-1}$ is directly implementable in hardware. We will choose not to implement this conversion between equivalent carry-save numbers, but simply to accept any carry-save representation produced by our circuit. Although the original specification was a general relation, we are happy to implement a function that gives only one of the possible outputs for any given input. The term $csv_{n-1}\,;\,csv_{n-1}^{-1}$ describes exactly what this choice entails.

In the design of a real circuit there would be two final steps: first the wiring should be rearranged to make the circuit more regular, and more suitable for implementation in hardware; next the combinational circuit should be pipelined by placing latches on internal arcs.

Pipelining can be done in the same way using behaviour-preserving transformation of the design. The relations that model combinational circuits are lifted to operate pointwise on streams of data values and a single extra primitive – the unit delay element – is introduced. The laws that presented here for 'combinational' relations hold also for relations on streams [Jones88]. In particular, the skewing theorem and its derivatives can be used to calculate the effects of placing delay elements on the internal arcs of linear arrays. A triangle of delay elements corresponds to the 'time-skewing' of a list of signals that is often necessary when arrays are pipelined. We first developed skewing theorems as a way of reasoning about the design of systolic arrays. Reference [Sheer88] uses relations to describe and reason about retiming and slowdown, the two main temporal transformations used in regular array design.

Some recent work on the design and verification of multipliers can be found in references [Borr89, Chin89, Luk88, Simon89].

## Conclusion

Describing both abstraction relations and circuits in a uniform relational framework works well. This is also demonstrated in reference [Mel89] which presents various kinds of abstraction for hardware verification using *Higher Order Logic*, which is also relational.

That we can use relations and their inverses allows us to start with a simple specification of a circuit as a representation changer. We then use a small set of algebraic laws to transform the specification into an implementation made of wiring and circuit primitives. The primitives are themselves small-scale representation changers. Some circuits, for example BCD to binary converters, can very obviously be described as representation changers. Others like the multiplier example above, do not immediately appear to be representation changers, but can be expressed as such when we see that multiplication is just as good an abstraction as addition.

Even in the simple examples presented here we have gone beyond the use of relations as a convenient way of describing networks of functions. We make essential use of the generality of relations.

In this paper we have only considered the binary and carry-save abstractions onto the naturals. It seems likely that our techniques will extend to other representations such as the negabinary representation of the integers as ladders of elements of $\{-1,0,1\}$. Another avenue of research opens up when we consider that abstractions need not be combinational. Once we have introduced streams and a delay element, we can describe not only how an abstract value is represented but also how the concrete value is presented to the circuit over time. In this way it should be possible to describe and design circuits with complex temporal behaviour – for example bit-serial circuits – using the techniques presented here. We plan to tackle more examples, involving more complicated representations.

ACKNOWLEDGEMENTS  This research is supported by SERC grant number GR/F 28939 (Relational Programming) and by SERC/IED grant number IED2/1/1759 (High Performance VLSI DSP Architectures). Sheeran gratefully acknowledges the support of a Royal Society of Edinburgh BP Research Fellowship. Thanks to Wayne Luk, Joe Morris, David Murphy and Lars Rossen for comments on earlier drafts.

# References

[deBakk86] J. W. de Bakker et al (eds.), *Mathematics and computer science*, North-Holland, 1986.

[Bhan89] A. S. Bhandal, V. Considine and G. E. Dixon, *An array processor for video picture motion estimation*, in [McCan89].

[Bird87] R. S. Bird, *An introduction to the theory of lists*, in [Broy87]. pp. 3–42. (Programming Research Group technical monograph PRG–56)

[Bird88] R. S. Bird, *Lectures on constructive functional programming*, in [Broy89]. (Programming Research Group technical monograph PRG–69)

[Borr89] D. Borrione and A. Salem, *Proving an on-line multiplier with OBJ and TACHE: a practical experience*, in [Claes89].

[Broy87] M. Broy (ed.), *Logic of programming and calculi of discrete design*, NATO advanced study institutes, Series F: Computer and systems sciences, Springer-Verlag, 1987.

[Broy89] M. Broy (ed.), *Constructive methods in computing science*, NATO advanced study institutes, Series F: Computer and systems sciences, Springer-Verlag, 1989.

[Chin89] S-K. Chin, *Verified synthesis functions for negabinary arithmetic hardware*, in [Claes89].

[Claes89] L. J. M. Claesen, *Applied formal methods for correct VLSI design*, North-Holland, 1989.

[Davis89] K. Davis and J. Hughes (eds.), *Functional programming, Glasgow 1989*, Springer Workshops in Computing, 1990.

[Jones86] G. Jones and W. Luk, *Exploring designs by circuit transformation*, in [Moore86]. pp. 91–98.

[Jones88] G. Jones and M. Sheeran, *Timeless truths about sequential circuits*, in [Tewk88]. pp. 245–259.

[Jones89] G. Jones, *Deriving the fast Fourier algorithm by calculation*, in [Davis89]. (Programming Research Group technical report PRG–TR–4–89)

[Jones90] G. Jones, *Designing circuits by calculation*, PRG technical report PRG–TR–10–90.

[Jouan85] J-P. Jouannaud (ed.), *Functional programming languages and computer architecture*, Springer LNCS 201, 1985.

[Luk88] W. Luk and G. Jones, *From specification to parametrised architectures*, in [Milne88]. pp. 267–288.

[Luk89]  W. Luk, G. Jones and M. Sheeran, *Computer-based tools for regular array design*, in [McCan89]. pp. 589–598.

[McCan82]  J. V. McCanny and J. G. McWhirter, *Completely iterative, pipelined multiplier array suitable for VLSI*, IEE Proc. Vol. 129, Pt. G, No. 2, April 1982. pp. 40–46.

[McCan89]  J. McCanny, J. McWhirter and E. Schwartzlander (eds.), *Systolic array processors*, Prentice Hall, 1989.

[Meert86]  L. G. L. T. Meertens, *Algorithmics – towards programming as a mathematical activity*, in [deBakk86].

[Meert89]  L. G. L. T. Meertens, *Constructing a calculus of programs*, in [Snep89].

[Mel89]  T. F. Melham, *Formalizing abstraction mechanisms for hardware verification in higher order logic*, Ph.D. Dissertation, University of Cambridge, August 1989.

[Milne88]  G. J. Milne (ed.), *The fusion of hardware design and verification*, North-Holland, 1988.

[Moore86]  W. Moore, A. McCabe and R. Urquhart (eds.), *Systolic arrays*, Adam Hilger, Bristol, 1986.

[Proeb87]  W. E. Proebster and H. Reiner (eds.), *Proc IEEE Comp Euro 87: VLSI and computers*, Hamburg, May 1987.

[Sheer84]  M. Sheeran, *muFP an algebraic VLSI design language*, in [Steel84]. pp. 104–112.

[Sheer85]  M. Sheeran, *Designing regular array architectures using higher-order functions*, in [Jouan85]. pp. 220–237.

[Sheer87]  M. Sheeran and G. Jones, *Relations + higher-order functions = hardware descriptions*, in [Proeb87]. pp. 303–306. (University of Glasgow Department of Computing Science technical report 87/R1)

[Sheer88]  M. Sheeran, *Retiming and slowdown in Ruby*, in [Milne88]. pp. 289–308.

[Simon89]  M. Simonis, *Formal verification of multipliers* in [Claes89].

[Snep89]  J. L. A. van de Snepscheut (ed.), *Mathematics of program construction*, Springer-Verlag LNCS 375, 1989.

[Steel84]  G. L. Steele Jr. et al (eds.), *Proc ACM Symp on LISP and functional programming*, 1984.

[Tewk88]  S. K. Tewksbury, B. W. Dickinson and S. C. Schwartz (eds.), *Concurrent computations: algorithms, architecture and technology*, Plenum Press, New York, 1988.

# George

In this workshop we have seen refinement methods that involve imperative, functional, and concurrent programs. Obviously, there are many advantages to be had from descriptions couched in these various styles, not least of all: functional descriptions are easy to write and to manipulate; and imperative and concurrent programs have many opportunities for implementation efficiency improvements. One goal of current research is to establish connections between these different styles and thereby exploit their benefits where they are most useful.

The next paper draws all three styles together. It uses a case study to illustrate a development style in which an applicative specification may be transformed into an imperative one. Moreover, concurrency may be introduced first with imperative combinators, but with details of processes specified in a functional style. The final implementation can describe all their algorithmic details imperatively. Interestingly, the technique described in the following paper also permits *event refinement:* atomic interactions may be refined into sequences of synchronisations at a lower level.

# Specifying and refining concurrent systems—an example from the RAISE project

## C. W. George and R. E. Milne [*]

**Abstract**

This paper illustrates a certain style of specification and and a certain notion of refinement. These permit the refinement of applicative specifications into imperative ones which express sequence and concurrency, the use of imperative specifications which are strongly "object-oriented" in that they do not even mention variables and channels, and the refinement of one apparently indivisible event into many such events. They are illustrated by three specifications, together with parts of the proofs that the second specification refines the first and the third specification refines the second.

## 1 Introduction

### 1.1 Intention

The RAISE project is an ESPRIT collaboration between Computer Resources International and Asea Brown Boveri in Denmark and International Computers Ltd and STC Technology Ltd in the United Kingdom. The project involves devising, trying out and revising a specification language, method and toolset which build on the prior experience of the partners (particularly with VDM) and allow for the treatment of modularity, "algebraic" data types, and concurrency.

In this paper we illustrate a style of specification and the notion of refinement adopted by the RAISE project. It arises from a trial of the RAISE language and method. This trial concerned the architecture of distributed processors with shared store. It needed an abstract enough architecture to allow unknown future implementations; it therefore required the use of a language and method which (among other things) supported concurrency, allowed imperative specifications to be independent of storage and communication techniques, and permitted single synchronisations at one level of refinement to be interpreted as multiple synchronisations at the next level. Because of this, the RAISE language and method were natural choices.

Here, in order to present the specifications and refinements in a short paper, we abstract away from the notion of storage and reduce the amount of concurrency involved in the original discussion [4]. We aim to demonstrate three important capabilities of the RAISE language and method in specifying and refining a system which is eventually implemented as a collection of communicating processes. These are:

- that the first specification of such a system can be an applicative one which is subsequently refined into an imperative one;

- that an imperative specification can describe the interactions between concurrent processes without selecting particular storage and communication techniques (or even the particular variables and channels needed by the techniques);

---

[*]STC Technology Ltd, London Road, Harlow, Essex, CM17 9NA, United Kingdom. This work was supported by STC Technology Ltd and the Commission of the European Communities under the ESPRIT programme, project number 315 (RAISE, "Rigorous Approach to Industrial Software Engineering"). It originated in work done by one of us (REM) for International Computers Ltd during a trial of the RAISE language and method.

- that the interactions between concurrent processes can be refined so that one apparently indivisible event is decomposed into many.

## 1.2 Outline

In section 2 there is an account of the RAISE language and method. In section 3 there is a description of the problem and a specification of various types and a value that are used in subsequent versions of the system. There then follow three sections. The first of these provides an initial specification of the system; each of the following two sections contains a version of the system that refines the one in the preceding section. For each version there is a brief description, the requirements in English, a specification in the RAISE language, a commentary on the RAISE notations, and, where appropriate, an outline of how to justify the claim that this version refines the preceding one. (To avoid irrelevant explanations, some minor liberties are taken with the language syntax.)

The first of these versions of the system, in section 4, relies on conditional equations between applicative functions. The second, in section 5, uses imperative combinators for concurrency but expresses the properties of processes in terms of the functions that interact with them, not in terms of programs that algorithmically implement them. The third, in section 6, implements the processes as programs by turning the apparently indivisible interactions into sequences of synchronised communications; these communications actually use the communication primitives of the RAISE language but could use other interaction functions with some other, equationally specified, communication protocol. (There is actually a gap in the second proof of refinement which needs to be filled by introducing material; this material is irrelevant to the central aim of this paper but can be found in the full version [4].)

# 2 Background

## 2.1 The style of specification

The RAISE language permits systems to be specified, designed and implemented (as executable software) in a single language. Accordingly, it provides a common conceptual framework for specification, design and implementation, and it facilitates proof of refinement. It allows both applicative specifications and imperative specifications; the imperative ones can be formulated and developed in much the same way as the applicative ones. It provides combinators for sequencing, concurrency, assignment to variables, and communication along channels; the concurrency and communication concepts are mainly those of CCS [2, 5] (with testing equivalence and value-passing) though the notation is closer to that of CSP [1, 3]. It allows related definitions and axioms to be classed together: members of the resulting (parameterised) "classes" are "objects", and may be declared as such (though there is no means of creating names for objects dynamically).

A specification in the RAISE language may be "model-oriented" (as in Z or VDM) or "property-oriented" (as in OBJ, rather loosely). A model-oriented specification can fit better with activities such as prototyping which let a user develop a feel for a specification. A property-oriented specification can offer more possibilities for refinement and allow a sharper focus on the essentials of the problem. A model-oriented specification can be longer or shorter than a property-oriented one, depending on the complexity of the control information and data representation. (Of course, once the types declared in one class are used to define types in another class, then the specification is in certain respects model-oriented even if the declarations of the original types are property-oriented.)

Fragments of specifications may be described by giving post-conditions or by stating equivalences between expressions. Both post-conditions and equivalences may be accompanied by pre-conditions, which restrict the sets of states in which the post-conditions and equivalences must hold. The post-conditions for the RAISE language can be defined in terms of equivalences. However, we do not exploit these post-conditions in this paper, as they are of limited help in dealing with concurrency; instead we rely on equivalences (sometimes with pre-conditions). Provided that they hold for all states, the equivalences are even congruences, which let one expression be substituted for another. This emphasis on congruences points to the use of a particular refinement relation, described in 2.2.

Functions in the RAISE language may be specified as being either partial or total; a total function does not diverge or deadlock (at least before it indulges in some communication).

## 2.2 The notion of refinement

The RAISE method provides guidelines for formulating and developing specifications and proof rules for demonstrating that specifications have particular formal properties. The proof rules may be needed either when validating a specification by exploring its properties (as in symbolic execution) or when verifying that one specification is related to another in a certain way. The relation between specifications that is central to this paper is the refinement relation. This refinement relation involves the removal of under-specification, in a sense that is clarified below. It should be distinguished from the relation used in some work on CSP, which involves the removal of non-determinism.

In the RAISE method a specification is the theory comprising all the consequences derivable from a class of definitions and axioms by using certain proof rules. The class itself is, strictly, a "presentation" of the theory; however, often we use the term "specification" to refer to a class presenting the theory, instead of to a theory. One specification is refined by another if all its consequences are consequences of the other specification (which may have further consequences as well). In other words, one class is refined by another if its theory is extended by the theory of the other class.

An axiom is typically either an expression with a post-condition or an equivalence between two expressions, perhaps with a pre-condition and some quantifiers. When one specification is refined by another its axioms must be consequences of the new specification. For instance, a specification that has the axiom

$$f() \,\textbf{post}\, (x = 0 \vee x = 1)$$

(in which the value of $x$ is obtained "in the state after the execution of $f()$") is refined by a specification that has the axiom

$$f() \,\textbf{post}\, x = 0$$

instead, because

$$(f() \,\textbf{post}\, (x = 0 \vee x = 1)) \Rightarrow (f() \,\textbf{post}\, x = 0)$$

By contrast, a specification that has the axiom

$$f() \equiv (x := 0 \sqcap x := 1)$$

(in which $\sqcap$ signifies an infixed internal choice combinator, as in CSP) is not generally refined by a specification that has the axiom

$$f() \equiv x := 0$$

instead.

Hence the removal of non-determinism from an expression, as in the last example above, does not necessarily entail the removal of under-specification from the specification in which the expression is embedded; in other words, removing non-determinism need not give rise to RAISE refinement. Nonetheless, in one important case, the removal of non-determinism from a RAISE specification entails the removal of under-specification. This case is where all the axioms in the specification take the form

$$\textbf{let}\, b = e \,\textbf{in}\, p \,\textbf{end} \equiv \textbf{let}\, b = e \,\textbf{in}\, \textbf{true}\, \textbf{end}$$

in which the post-condition *p* is monotonic in the result *b* and store of the expression *e* (or where the result and store are first-order). (This form of axiom actually provides a "weak" post-condition, which can be satisfied by an expression that is not terminating or deterministic.) However, not all axioms, particularly for concurrent systems, can be cast conveniently in this form.

We must also mention a further important distinction, between an equivalence between the effects of expressions and an equality between the results returned by expressions. Except for specifications in which every expression is applicative, terminating and deterministic these two notions are different. In the RAISE language, the relation $\equiv$ expresses an equivalence between the effects of expressions when both the expressions are executed starting in a particular state; the relation $=$ expresses an equality between the results returned by expressions when the expressions are executed sequentially starting in a particular state. For instance,

$$(x := x + 1 \; ; \; x) \equiv (x := x + 1 \; ; \; x)$$

holds but

$$(x := x + 1 \; ; \; x) = (x := x + 1 \; ; \; x)$$

does not hold (because, remember, there will be two successive assignments to *x* in the execution), whereas

$$(x := 0 \; ; \; 0) \equiv (x := 1 \; ; \; 0)$$

does not hold but

$$(x := 0 \; ; \; 0) = (x := 1 \; ; \; 0)$$

holds.

## 3 The common basis for the specifications

### 3.1 Description

The system under consideration in this paper manages the sharing of "facilities" between "tasks". We are not concerned with what facilities are — they might be the storage locations of the system in the original trial, hardware devices required by programs, or even both programs and their devices. Tasks represent potentially concurrent operations (though in this paper we shall omit the axioms which describe how such operations appear to interleave).

Users may "begin" a task by reserving a set of facilities provided they are not already available to some other task. They may "do" a task with a sequence of facilities provided all facilities in the sequence have been reserved by a prior "begin". They may "end" a task by releasing all the facilities reserved for it.

### 3.2 Types

#### 3.2.1 Requirements

Two types (*Task* and *Facility*) are used frequently but can be left abstract: nothing need be said about them other than that they exist (which provides the right to use tests for equality between their members).

The ultimate implementation makes use of the set of all tasks in ways which require it to be finite (to avoid unbounded non-determinism). For convenience, the fact that it is finite is postulated here.

    *Types* =
      **class**
        **type**
          *Task*,

         *Facility*
      **value**
         *all_tasks* : *Task*-**set** = {*t* | *t* : *Task*}
   **end**


### 3.2.2 Commentary

*Types* is a class expression containing just a type declaration and a value declaration. The type declaration provides merely names for the two types. The value declaration provides a name for the set of all tasks and also indicates that the set of tasks is finite (by the use of -**set** instead of -**infset**).

The functions for beginning, doing and ending tasks are not include in the common basis, because their signatures depend on a type which is declared differently in each specification.

## 4 The first specification

### 4.1 Description

The system allows the beginning, doing and ending of tasks. To begin a task one must identify the task and the set of facilities required by it. To do a task one must identify the task and the sequence of facilities required. To end a task one need only identify the task. The system has a current "status" which can be queried about the facilities currently unavailable to a task (because they are available to other tasks) and about the facilities available to a task.

### 4.2 Operations

#### 4.2.1 Requirements

The system must satisfy the following constraints:

- no facility should ever be available to more than one task;

- if a task has no facilities available to it already and the facilities that it wants are not unavailable to it already, it can be begun, whereupon the facilities that it wants become available to it but the availability of facilities does not otherwise change;

- if a task has the facilities that it wants available to it already, it can be done, whereupon the availability of facilities does not change;

- a task can be ended, whereupon the facilities that are available to it cease to be available to it but the availability of facilities does not otherwise change.

#### 4.2.2 Specification

   $Operations_1(X : Types) =$
      **class**
         **type**
            *Act*
         **value**
            *available* : *Act* $\to$ *X*.*Task* $\to$ *X*.*Facility*-**set**,
            *unavailable* : *Act* $\to$ *X*.*Task* $\to$ *X*.*Facility*-**set**
            *unavailable*(*a*)(*t*) $\equiv$
               {*f* | *f* : *X*.*Facility* • $\exists t_1$ : *X*.*Task* • $t \neq t_1 \land f \in$ *available*(*a*)($t_1$)}
         **axiom**
            /* no facility is available to more than one task */

$\forall\, a : Act \bullet \forall\, t : X.Task \bullet available(a)(t) \cap unavailable(a)(t) = \{\}$
**value**
/* begin */
$b : (X.Task \times X.Facility\text{-set} \times Act) \xrightarrow{\sim} Act,$
/* do */
$d : (X.Task \times X.Facility\text{-list} \times Act) \xrightarrow{\sim} Act,$
/* end */
$e : (X.Task \times Act) \xrightarrow{\sim} Act$
**axiom**
/* tasks may be begun if no facilities are available to them already */
/* and the wanted facilities are not unavailable to them already */
$\forall\, t : X.Task, f\_s : X.Facility\text{-set}, a : Act \bullet$
  $(available(a)(t) = \{\} \wedge f\_s \cap unavailable(a)(t) = \{\}) \Rightarrow$
    $available(b(t, f\_s, a)) = \lambda\, t_1 \bullet \textbf{if } t_1 = t \textbf{ then } f\_s \textbf{ else } available(a)(t_1) \textbf{ end},$
/* tasks may be done if the wanted facilities are available to them already */
$\forall\, t : X.Task, f\_l : X.Facility\text{-list}, a : Act \bullet$
  $\textbf{elems}\, f\_l \subseteq available(a)(t) \Rightarrow$
    $available(d(t, f\_l, a)) = available(a),$
/* tasks may be ended */
$\forall\, t : X.Task, a : Act \bullet$
  $available(e(t, a)) = \lambda\, t_1 \bullet \textbf{if } t_1 = t \textbf{ then } \{\} \textbf{ else } available(a)(t_1) \textbf{ end}$
**end**

### 4.2.3 Commentary

*Types* is made a parameter to *Operations$_1$*; it can later be instantiated with some more definite notion of what tasks and facilities are.

Basic class expressions in the RAISE language consist of a sequence of declarations between the words **class** and **end**. Declarations may be of types, values, axioms (all illustrated here), variables, channels, schemes and objects.

Types may be abstract like *Act* (or like *Task* and *Facility* defined in *Types*). They may also be given more concrete definitions, as we shall see later.

Values (which include functions) must be given types. They may also be given definitions (like *unavailable*) or not (like *available*, *b*, *d* and *e*). The values may be constrained by axioms, as here. Axioms are taken to hold for all states, so axioms which consist of equivalences between expressions provide congruences: one of the expressions may be substituted for the other.

The types of functions will include an arrow, either $\rightarrow$ or $\xrightarrow{\sim}$ . The first of these indicates a total function, the second a partial one (which may be refined into a total one). A total function does not diverge or deadlock before either waiting to communicate or terminating.

The type *Act* may be regarded as providing the status of the system. The functions *available* and *unavailable* query the status to see what facilities are currently available or not to the tasks. They are both total.

Beginning, doing and ending tasks (abbreviated to *b*, *d* and *e* respectively) are all applicative, partial functions. The axioms governing them, together with the axioms relating *available* and *unavailable*, capture the requirements very directly.

## 4.3 Justification

In order to establish that a top-level specification captures the intended intuitions that a developer would have, it is usual to prove some theorems. These theorems should show that the specification itself has certain desirable properties (such as the property of being refinable into a particular potential implementation).

For the specification in 4.2.2 the axioms are given in such a general form that several theorems which one might wish to prove about them are immediate consequences. For example, we can check that beginning, doing and ending a task where the facilities used are all properly reserved will result in some suitable final status. We can do this by evaluating the expression

$available(e(t, d(t, l, b(t, \mathsf{elems}\, l, a))))$

where $available(a)(t) = \{\} \wedge \mathsf{elems}\, l \cap unavailable(a)(t) = \{\}$. To do this we use the axioms for $e$, $d$ and $b$ in turn, as follows:

$available(e(t, d(t, l, b(t, \mathsf{elems}\, l, a))))$
=
$\lambda t_1 \bullet \mathsf{if}\ t_1 = t\ \mathsf{then}\ \{\}\ \mathsf{else}\ available(d(t, l, b(t, \mathsf{elems}\, l, a)))(t_1)\ \mathsf{end}$ axiom for $e$
=
$\lambda t_1 \bullet \mathsf{if}\ t_1 = t\ \mathsf{then}\ \{\}\ \mathsf{else}\ available(b(t, \mathsf{elems}\, l, a))(t_1)\ \mathsf{end}$ axiom for $d$
=
$\lambda t_1 \bullet \mathsf{if}\ t_1 = t\ \mathsf{then}\ \{\}$
$\quad\quad \mathsf{else}\ (\lambda t_1 \bullet \mathsf{if}\ t_1 = t\ \mathsf{then}\ \mathsf{elems}\, l\ \mathsf{else}\ available(a)(t_1)\ \mathsf{end})(t_1)$
$\quad \mathsf{end}$ axiom for $b$
=
$\lambda t_1 \bullet \mathsf{if}\ t_1 = t\ \mathsf{then}\ \{\}$
$\quad\quad \mathsf{else\ if}\ t_1 = t\ \mathsf{then}\ \mathsf{elems}\, l\ \mathsf{else}\ available(a)(t_1)\ \mathsf{end}$
$\quad \mathsf{end}$
=
$\lambda t_1 \bullet \mathsf{if}\ t_1 = t\ \mathsf{then}\ \{\}\ \mathsf{else}\ available(a)(t_1)\ \mathsf{end}$
=
$available(a)$

This is what we should expect — the initial and final statuses have the same facilities available to all tasks.

We can also check some other properties, such as can we have more than one 'do' for a particular task? (The answer is "yes"). Does it matter if we have more than one 'end' for a task? (The answer is "no"). We could also investigate the interleaving of different tasks.

If we ask what happens if two tasks try to reserve the same or overlapping sets of facilities then we get expressions that we cannot reduce because the condition guarding the axiom for $b$ for at least the second task will be false. Hence the answer is "don't know". In other words our specification is currently too weak to decide this issue; it may be resolved later and certainly will be resolved in the final implementation.

# 5 The second specification

## 5.1 Description

The initial refinement in this paper takes into account the wish to provide a concurrent implementation by taking *Act* to be a type of processes (which represent the possible activities of the system) and by taking $b$, $d$ and $e$ to be functions intended to interact with these processes.

## 5.2 Operations

### 5.2.1 Requirements

The system should not have a status whereby it may diverge or deadlock before either waiting to communicate or terminating. It ensures that no facilities are reserved by more than one task at once in accordance with rules like those in 4.2.1.

Beginning, doing and ending a task preserve liveness, in certain circumstances: when they are performed with a system making appropriate facilities available, they do not diverge or deadlock before terminating and they make the system continue to have a status according to which it does not diverge or deadlock before either waiting to communicate or terminating.

### 5.2.2 Specification

$Operations_2(X : Types) =$
**class**
  **value**
    /* begin */
    $b : (X.Task \times X.\text{Facility-set} \times Act) \xrightarrow{\sim} Act$
    $b(t, f\_s, a)() \equiv b\_x(t, f\_s) \mathbin{\|} a(),$
    /* do */
    $d : (X.Task \times X.\text{Facility-list} \times Act) \xrightarrow{\sim} Act$
    $d(t, f\_l, a)() \equiv d\_x(t, f\_l) \mathbin{\|} a(),$
    /* end */
    $e : (X.Task \times Act) \xrightarrow{\sim} Act$
    $e(t, a)() \equiv e\_x(t) \mathbin{\|} a()$
  **type**
    $Act = \{|a : \textbf{Unit} \rightarrow \textbf{write any in any out any Unit} \bullet acting(a)|\}$
  **value**
    $acting : (\textbf{Unit} \rightarrow \textbf{write any in any out any Unit}) \rightarrow \textbf{Bool},$
    $available : Act \rightarrow X.Task \rightarrow X.\text{Facility-set},$
    $unavailable : Act \rightarrow X.Task \rightarrow X.\text{Facility-set}$
    $unavailable(a)(t) \equiv$
      $\{f \mid f : X.Facility \bullet \exists t_1 : X.Task \bullet t \neq t_1 \land f \in available(a)(t_1)\}$
  **axiom**
    /* no facility is available to more than one task */
    $\forall a : Act \bullet \forall t : X.Task \bullet available(a)(t) \cap unavailable(a)(t) = \{\}$
  **value**
    /* begin */
    $b\_x : (X.Task \times X.\text{Facility-set}) \xrightarrow{\sim} \textbf{in any out any Unit},$
    /* do */
    $d\_x : (X.Task \times X.\text{Facility-list}) \xrightarrow{\sim} \textbf{in any out any Unit},$
    /* end */
    $e\_x : X.Task \xrightarrow{\sim} \textbf{in any out any Unit}$
  **axiom**
    /* tasks may be begun if no facilities are available to them already */
    /* and the wanted facilities are not unavailable to them already */
    $\forall t : X.Task, f\_s : X.\text{Facility-set}, a_1 : Act \bullet$
      $(available(a_1)(t) = \{\} \land f\_s \cap unavailable(a_1)(t) = \{\}) \Rightarrow$
      $\exists a_2 : Act \bullet$
        $(b\_x(t, f\_s) \mathbin{\|} a_1() \equiv a_2()) \land$
        $available(a_2) = \lambda t_1 \bullet \textbf{if } t_1 = t \textbf{ then } f\_s \textbf{ else } available(a_1)(t_1) \textbf{ end},$
    /* tasks may be done if the wanted facilities are available to them already */
    $\forall t : X.Task, f\_l : X.\text{Facility-list}, a_1 : Act \bullet$
      $\textbf{elems } f\_l \subseteq available(a_1)(t) \Rightarrow$
      $\exists a_2 : Act \bullet$
        $(d\_x(t, f\_l) \mathbin{\|} a_1() \equiv a_2()) \land$
        $available(a_2) = available(a_1),$
    /* tasks may be ended */
    $\forall t : X.Task, a_1 : Act \bullet$

$\exists\, a_2 : Act \bullet$
$(e\_x(t) \mathbin{\|\!\|} a_1() \equiv a_2()) \land$
$available(a_2) = \lambda\, t_1 \bullet \textbf{if}\ t_1 = t\ \textbf{then}\ \{\}\ \textbf{else}\ available(a_1)(t_1)\ \textbf{end}$
**end**

### 5.2.3 Commentary

The declaration of the type *Act* indicates that any function which is a member of this type must take a parameter with type **Unit**, can write to (or read from) any variable, can input from any channel, can output to any channel, and can only return a result which has type **Unit**. (Here **Unit** is a type containing only the single value '()'.) Any process of type *Act* must also satisfy the predicate *acting*. We will specify *acting* in more detail in the next refinement in order to constrain possible implementations further; it may be said to model an (underspecified) "invariant" on the possible statuses provided by the type *Act*.

The types of the functions $b\_x$, $d\_x$ and $e\_x$ indicate that they do not write to (or read from) any variable, can input from any channel and can output from any channel. (In more complicated problems than this it is necessary to allow these functions to write to variables in objects other than those accessed by the activities of the system.)

The interaction between the system and these functions is specified in terms of the interlocking combinator, $\mathbin{\|\!\|}$, rather than the concurrency combinator, $\|$, because the intention is to localise the description of any possible communication between between the processes. The concurrency combinator, $\|$, allows input and output to be either synchronised or interleaved; it thereby permits the occurrence of arbitrarily long sequences of communications which are left outstanding before or during the interaction. The interlocking combinator, $\mathbin{\|\!\|}$, requires input and output to be synchronised; it thereby prevents the occurrence of such sequences of communications. Both combinators give rise to a non-deterministic choice between legitimate synchronisations between inputs and outputs, but $\|$ also exposes outstanding possibilities for communication whilst $\mathbin{\|\!\|}$ conceals them. (In more complicated problems than this it is necessary to indicate in the interlocking combinator which communications need to be synchronised, as described in the full version of this paper [4].)

## 5.3 Justification

We want to show that the second specification refines the first, or, in other words, that for any object $X$ in the class *Types* the theory of the second class extends that of the first. To do this we must do the following two things:

- check that the signature of the second class extends that of the first, in that for every definition in the first class there is a corresponding definition in the second having the same kinds (**type**, **value** or whatever) and, for values, variables and channels, the same type (except where the second class interprets an abstract type set up in the first class);

- prove that the axioms of the first class, together with any properties implied by the definitions, are implied by the axioms and definitions of the second.

The signature check is easy in this case. *Operations$_1$* exports the type *Act* and the values *unavailable*, *available*, *b*, *d* and *e*. *Operations$_2$* has the same parameter *Types* and exports the same names with the same kinds and (for the values) the same types.

To prove that the axioms of the first class, together with any properties implied by definitions, are implied by the axioms and definitions of the second is also straightforward. We note firstly that the definition of *unavailable* is repeated in the second, and so is the axiom ensuring disjointness of the sets of available facilities. We are then left with the axioms for *b*, *d* and *e*. To take the axiom *b*, we have to show in *Operations$_2$* that

$\forall\, t : X.Task, f\_s : X.Facility\text{-}\mathbf{set}, a : Act \bullet$
$(available(a)(t) = \{\} \land f\_s \cap unavailable(a)(t) = \{\}) \Rightarrow$

$$available(b(t, f\_s, a)) = \lambda\, t_1 \bullet \text{if } t_1 = t \text{ then } f\_s \text{ else } available(a)(t_1) \text{ end}$$

To show this we note that in $Operations_2$ we have

$b(t, f\_s, a)()$
$\equiv$
$\quad b\_x(t, f\_s) \mathbin{\|} a()$ \hfill definition of $b$
$\equiv$
$\quad a_2()$
for some $a_2$, provided $available(a)(t) = \{\} \wedge f\_s \cap unavailable(a)(t) = \{\}$ \hfill axiom for $b\_x$

and the rest follows immediately from the axiom for $b\_x$ The axioms for $d$ and $e$ are similarly easy to justify.

# 6 The third specification

## 6.1 Description

The other refinement in this paper does not change the relation between $b$, $d$ and $e$ (on the one hand) and $Act$ and so on (on the other hand). It does, however, provide explicit programs for the activities of the system and for the functions which interact directly with those programs ($b\_x$, $d\_x$ and $e\_x$).

The functions $b\_x$ and $d\_x$ are decomposed into sequences of communications. In the presence of axioms about the concurrent processing of tasks, as in the full paper [4], this decomposition amounts to refining single apparently indivisible operations into multiple operations.

## 6.2 Operations

### 6.2.1 Requirements

The system must provide concrete algorithms that maintain the axioms of 5.2.2. To achieve this it relies on what is probably the simplest scheduling algorithm, in which only one task is allowed to be active at any one time. Hence once one task has been begun no other task will be allowed to begin until it has been ended. More complicated protocols could be adopted, such as allowing tasks to begin but queueing them if they could interfere with active tasks until these have finished, but this would complicate the specification and the justification.

### 6.2.2 Specification

$Operations_3(X : Types) =$
 **class**
  **value**
   /* begin */
   $b : (X.Task \times X.Facility\text{-set} \times Act) \xrightarrow{\sim} Act$
   $b(t, f\_s, a)() \equiv b\_x(t, f\_s) \mathbin{\|} a(),$
   /* do */
   $d : (X.Task \times X.Facility\text{-list} \times Act) \xrightarrow{\sim} Act$
   $d(t, f\_l, a)() \equiv d\_x(t, f\_l) \mathbin{\|} a(),$
   /* end */
   $e : (X.Task \times Act) \xrightarrow{\sim} Act$
   $e(t, a)() \equiv e\_x(t) \mathbin{\|} a()$
  **type**
   $Act = \{|a : \textbf{Unit} \rightarrow \textbf{write any in any out any Unit} \bullet acting(a)|\}$
  **value**

$acting : (\textbf{Unit} \rightarrow \textbf{write any in any out any Unit}) \xrightarrow{\sim} \textbf{Bool}$
$acting(a) \equiv$
  $\exists! \, t\_s_1 : X.Task\text{-set}, t\_s_2 : X.Task\text{-set}, f\_s : X.Facility\text{-set} \bullet$
    $is\_proper(t\_s_1, t\_s_2, f\_s) \land$
    $a = (\lambda() \bullet may\_start := t\_s_1 \,;\, may\_do := t\_s_2 \,;\, reservation := f\_s \,;\, act()),$
$available : Act \rightarrow X.Task \rightarrow X.Facility\text{-set}$
$available(a)(t) \equiv$
  $\textbf{let } t\_s_1 : X.Task\text{-set}, t\_s_2 : X.Task\text{-set}, f\_s : X.Facility\text{-set} \bullet$
    $is\_proper(t\_s_1, t\_s_2, f\_s) \land$
    $a = (\lambda() \bullet may\_start := t\_s_1 \,;\, may\_do := t\_s_2 \,;\, reservation := f\_s \,;\, act())$
  $\textbf{in if } t \in t\_s_2 \textbf{ then } f\_s \textbf{ else } \{\} \textbf{ end end},$
$unavailable : Act \rightarrow X.Task \rightarrow X.Facility\text{-set}$
$unavailable(a)(t) \equiv$
  $\{f \mid f : X.Facility \bullet \exists t_1 : X.Task \bullet t \neq t_1 \land f \in available(a)(t_1)\}$

**variable**
  $reservation : X.Facility\text{-set} := \{\},$
  $may\_start : X.Task\text{-set} := X.all\_tasks,$
  $may\_do : X.Task\text{-set} := \{\}$

**channel**
  $b\_f\_s : X.Facility\text{-set},$
  $d\_f\_l : X.Facility\text{-list}$

**value**
  $is\_proper : (X.Task\text{-set} \times X.Task\text{-set} \times X.Facility\text{-set}) \rightarrow \textbf{Bool}$
  $is\_proper(may\_start, may\_do, reservation) \equiv$
    $(may\_start = \{\} \land \textbf{card } may\_do = 1) \lor$
    $(may\_start = X.all\_tasks \land may\_do = \{\} \land reservation = \{\}),$
  $act : \textbf{Unit} \rightarrow \textbf{write any in any out any Unit}$
  $act() \equiv$
    $\textbf{while true do } body() \textbf{ end},$
  $body : \textbf{Unit} \rightarrow \textbf{write any in any out any Unit}$
  $body() \equiv$
    $[] \, \{U[t].b\,!() \,;\, reservation := b\_f\_s\,? \,;\, may\_start := \{\} \,;\, may\_do := \{t\} \mid$
       $t : X.Task \bullet t \in may\_start \lor t \in may\_do\} \, []$
    $[] \, \{U[t].d\,!() \,;\, perform(d\_f\_l\,?) \mid t : X.Task \bullet t \in may\_do\} \, []$
    $[] \, \{U[t].e\,!() \,;\, may\_start := X.all\_tasks \,;\, may\_do := \{\} \,;\, reservation := \{\} \mid$
       $t : X.Task\},$
  $perform : X.Facility\text{-list} \rightarrow \textbf{write any Unit}$

**axiom**
  $\forall \, t\_s_1 : X.Task\text{-set}, t\_s_2 : X.Task\text{-set}, f\_s : X.Facility\text{-set}, f\_l : X.Facility\text{-list} \bullet$
    $is\_proper(t\_s_1, t\_s_2, f\_s) \land \textbf{elems } f\_l \subseteq f\_s \Rightarrow$
    $(may\_start := t\_s_1 \,;\, may\_do := t\_s_2 \,;\, reservation := f\_s \,;\, perform(f\_l) \equiv$
     $perform(f\_l) \,;\, may\_start := t\_s_1 \,;\, may\_do := t\_s_2 \,;\, reservation := f\_s)$

**value**
  /* begin */
  $b\_x : (X.Task \times X.Facility\text{-set}) \xrightarrow{\sim} \textbf{in any out any Unit}$
  $b\_x(t, f\_s) \equiv U[t].b\,? \,;\, b\_f\_s\,!\,f\_s,$
  /* do */
  $d\_x : (X.Task \times X.Facility\text{-list}) \xrightarrow{\sim} \textbf{in any out any Unit}$
  $d\_x(t, f\_l) \equiv U[t].d\,? \,;\, d\_f\_l\,!\,f\_l,$
  /* end */
  $e\_x : X.Task \xrightarrow{\sim} \textbf{in any out any Unit}$
  $e\_x(t) \equiv U[t].e\,?$

```
      object
        U[t : X.Task] :
        class
          channel
            b : Unit,
            d : Unit,
            e : Unit
        end
end
```

### 6.2.3   Commentary

In 6.2.2 the functions $b$, $d$ and $e$ are related to the functions $b\_x$, $d\_x$ and $e\_x$ in the same way as in 5.2.2. It would be possible to avoid repeating the relation, by introducing appropriate parameters for the classes, but we have not done so, to avoid naming extra classes.

Though the definitions of *Act* and *unavailable* in 6.2.2 are just as in 5.2.2, the definitions of *acting* and *available* now depend on the status of the system. In particular, the constraint imposed by *acting* ensures that any member of the type *Act* is definable by a unique assignment to the variables of a 'proper' collection of values followed by the process *act*. This constraint is exploited in the definition of *available*.

The status of the system is held in three variables. *reservation* contains the set of facilities reserved by the currently started task. *may_start* contains the set of task values for which only a 'begin' or 'end' will be accepted. *may_do* contains the set of task values for which a 'begin', 'do' or 'end' will be accepted. These variables are all given initial values so that initially any task may begin or end and none may do. The initial reservation is empty.

The channels $b\_f\_s$ and $d\_f\_s$ are to pass the sets and lists of facilities to the system.

*is_proper* is an auxiliary function intended to encapsulate the contents of variables which characterise a 'well-formed' status.

The process *act* loops for ever waiting to synchronise with a user process on one of the channels $U[t].b$, $U[t].d$ or $U[t].e$ according to the function *body* (which is introduced merely for convenience in 6.3). The expression

$$e_1 \,[]\, e_2$$

is the external choice between $e_1$ and $e_2$. The expression

$$[]\, \{e \mid b : t \bullet p\}$$

is the external choice between the expressions in the set $\{e \mid b : t \bullet p\}$. If the set is empty this external choice reduces to **stop**, which is such that

$$e \,[]\, \textbf{stop} \equiv e$$

holds for every expression $e$. Hence the possible choices of $t$ for which *act* will synchronise, and the choice of which of these channels will then be accepted, are governed by the values in the state variables. When one of these synchronisations occur *act* may use the value passed in a second synchronisation — in the case of a 'begin' to reserve the facilities and in the case of a 'do' to perform the function by applying *perform*. This function is left underspecified except for the requirement that it does not interfere with the status of the system. (This requirement could be weakened to allow communication to take place during an invocation of *perform*.)

The functions $b\_x$, $d\_x$ and $e\_x$ each wait for their initial synchronisation and then (for the first two) output their parameter values.

$U$ is an array of objects indexed by $X.Task$. Each object in the array is a member of a particular class which defines three channels; each of these channels can commmunicate a value of type **Unit**. The channels defined in any object are among the channels encompassed by **any** (provided that the object is itself defined within the bounds of that **any**, which in practice means provided that the object is not a parameter to a class containing the **any**).

## 6.3 Justification

The check on the signature of 6.2.2 is trivial.

Moreover, the definitions of $b$, $d$ and $e$, *Act* and *unavailable* are the same in 5.2.2 as in 6.2.2. We are left to deal with the axioms for *available*, $b\_x$, $d\_x$ and $e\_x$.

We first present a justification in terms of 6.2.2 for the critical property that no two tasks may access the same facility, which is stated as

$\forall a : Act, t : X.Task \bullet$
  $available(a)(t) \cap unavailable(a)(t) = \{\}$

We note initially that $a()$ is uniquely represented by assignments followed by $act()$. If

$a = (\lambda() \bullet may\_start := t\_s_1 \ ; \ may\_do := t\_s_2 \ ; \ reservation := f\_s \ ; \ act())$

then $is\_proper(t\_s_1, t\_s_2, f\_s)$ and

$available(a)(t) = \textbf{if} \ t \in t\_s_2 \ \textbf{then} \ f\_s \ \textbf{else} \ \{\} \ \textbf{end}$ \hfill $(1)$

Now suppose there were some facility $f$ that were in both $available(a)(t)$ and in $unavailable(a)(t)$. Then from the definition of *unavailable* there would be a $t_1$ different from $t$ such that $f$ is in $available(a)(t_1)$. Yet from (1) this is impossible since $is\_proper$ implies that $t\_s_2$ is either empty or contains one task value. Accordingly the critical property holds.

We next present a justification of the axiom for $b\_x$ in 5.2.2. We must show that

$\forall t : X.Task, f\_s : X.Facility\text{-}set, a_1 : Act \bullet$
  $(available(a_1)(t) = \{\} \land f\_s \cap unavailable(a_1)(t) = \{\}) \Rightarrow$
  $\exists a_2 : Act \bullet$
    $(b\_x(t, f\_s) \ |\!|\!| \ a_1() \equiv a_2()) \land$
    $available(a_2) = \lambda \ t_1 \bullet \textbf{if} \ t_1 = t \ \textbf{then} \ f\_s \ \textbf{else} \ available(a_1)(t_1) \ \textbf{end}$

We proceed as in the first justification and assume that

$a_1 = (\lambda() \bullet may\_start := t\_s_1 \ ; \ may\_do := t\_s_2 \ ; \ reservation := f\_s \ ; \ act())$

for some 'proper' triple $(t\_s_1, t\_s_2, f\_s)$. Now we can see from $is\_proper$ that $t\_s_1$ is either empty or equal to $all\_tasks$. Now the first case can only occur if some other task has started, and this one will not be able to start until that other one has finished. Thus a full proof of this axiom is only possible by describing a notion of 'well-formed task' (a sequence of begin, do and end) and hence a notion of well-formed interleavings of such tasks. We can then show that if all the interleaved tasks are well-formed then any particular task will eventually be able to start. To keep this paper brief and comprehensible we do not deal with this problem in this paper; instead we simply assume that $t\_s_1$ is equal to $all\_tasks$. Then (from $is\_proper$) $t\_s_2$ and $f\_s$ must be empty, and we have

  $b\_x(t, f\_s) \ |\!|\!| \ a_1()$
$\equiv$
  $(U[t].b? \ ; \ b\_!f\_s) \ |\!|\!| \ (may\_start := X.all\_tasks \ ; \ may\_do := \{\} \ ; \ reservation := \{\} \ ; \ act())$
$\equiv$
  $may\_start := X.all\_tasks \ ; \ may\_do := \{\} \ ; \ reservation := \{\} \ ;$
  $(U[t].b? \ ; \ b\_!f\_s) \ |\!|\!| \ \textbf{while true do} \ body() \ \textbf{end}$
$\equiv$
  $may\_start := X.all\_tasks \ ; \ may\_do := \{\} \ ; \ reservation := \{\} \ ;$
  $(U[t].b? \ ; \ b\_!f\_s) \ |\!|\!| \ (body() \ ; \ \textbf{while true do} \ body() \ \textbf{end})$
$\equiv$

$$may\_start := X.all\_tasks \ ; \ may\_do := \{\} \ ; \ reservation := \{\} \ ;$$
$$(U[t].b? \ ; \ b\_!f\_s) \parallel\!\!\!\parallel (body() \ ; \ act())$$
$$\equiv$$
$$may\_start := X.all\_tasks \ ; \ may\_do := \{\} \ ; \ reservation := \{\} \ ; \qquad (2)$$
$$((U[t].b? \ ; \ b\_!f\_s) \parallel\!\!\!\parallel body()) \parallel\!\!\!\parallel act()$$
$$\equiv$$
$$may\_start := X.all\_tasks \ ; \ may\_do := \{\} \ ; \ reservation := \{\} \ ; \qquad (3)$$
$$(b\_f\_s!f\_s \parallel\!\!\!\parallel (reservation := b\_f\_s? \ ; \ may\_start := \{\} \ ; \ may\_do := \{t\})) \parallel\!\!\!\parallel act()$$
$$\equiv$$
$$may\_start := X.all\_tasks \ ; \ may\_do := \{\} \ ; \ reservation := \{\} \ ; \qquad (4)$$
$$reservation := f\_s \ ; \ may\_start := \{\} \ ; \ may\_do := \{t\} \ ; \ act()$$
$$\equiv$$
$$may\_start := \{\} \ ; \ may\_do := \{t\} \ ; \ reservation := f\_s \ ; \ act()$$
$$\equiv$$
$$a_2()$$

for some $a_2$ (noting that $is\_proper(\{\}, \{t\}, f\_s)$) and then, from (1)

$$available(a_2)(t_1) = \text{if } t_1 \in \{t\} \text{ then } f\_s \text{ else } \{\} \text{ end}$$
$$\equiv$$
$$available(a_2) = \lambda \, t_1 \bullet \text{if } t_1 = t \text{ then } f\_s \text{ else } \{\} \text{ end}$$

Step (2) above used the rule that

$$e_1 \parallel\!\!\!\parallel (e_2 \ ; \ e_3) \equiv (e_1 \parallel\!\!\!\parallel e_2) \parallel\!\!\!\parallel e_3$$

provided that $e_1 \parallel\!\!\!\parallel e_2$ does not deadlock or diverge and does not input or output on the same channels as $e_3$. This is typically true if the interlocking of $e_1$ and $e_2$ 'consumes' all the communications of $e_2$, which is true here, where $e_1 \parallel\!\!\!\parallel e_2$ eventually reduces to a sequence of assignments.

Steps (3) and (4) are based on the basic rule for relating external choice to interlocking, for which

$$((\text{let } b = c \, ? \, \text{in } e_1(b) \text{ end}) \, [] \, e_4) \parallel\!\!\!\parallel ((c!e_2 \ ; \ e_3) \, [] \, e_5)$$
$$\equiv$$
$$\text{let } b = e_2 \text{ in } e_1(b) \text{ end} \parallel\!\!\!\parallel e_3$$

provided that $e_2$ is read-only and convergent, that $e_4$ requires communication along channels other than $c$ and the channels of $e_5$, and that $e_5$ requires communication along channels other than $c$ and the channels of $e_4$.

# 7 References

[1] Brookes, S.D., Hoare, C.A.R., and Roscoe, A.W., *A Theory of Communicating Sequential Processes*, **Journal of the ACM 31**, 3 (July 1984), 560-599.

[2] De Nicola, R., and Hennessy, M., *CCS without $\tau$'s*, **Proceedings of the International Joint Conference on Theory and Practice of Software Development**, Lecture Notes in Computer Science 249 (Springer-Verlag, Berlin, 1985), 138-152.

[3] Hoare, C.A.R., **Communicating Sequential Processes** (Prentice-Hall, 1985).

[4] Milne, R.E., *Specifying and refining concurrent systems*, RAISE/STC/REM/13/V2, STC (1990).

[5] Milner, A.J.R.G., **A calculus of communicating systems**, Lecture Notes in Computer Science 92 (Springer-Verlag, 1980).

# Sennett

The use of the case study is a well-known and important research tool: it may uncover new research problems upon which to work, for instance. However, equally importantly, it may help in the evaluation of the practicality of using research results. For example, it can tell us whether our techniques—that may look so good on paper—actually scale up to tackle problems of a realistic size. Much of the work to date on refinement has concentrated on the ability to develop clever, intricate, but rather small, programs—"Swiss watch mechanisms". If such work is to have any impact on industrial programming practice, then it must be demonstrated that industrial-sized programs can be developed cost-effectively.

If completely formal developments of significant programs are beyond our capabilities at present, then how might our techniques be adapted to work informally, but with a high degree of assurance? In the following paper the author describes the application of a particular refinement technique, and addresses these issues.

# Using refinement to convince: lessons learned from a case study

C T Sennett[*]

**Abstract**

This is a discussion of a case study in classical data refinement. The specification is the non-trivial problem of pattern matchine in the programming language ML. The case study illustrates how refinement can assist in the design process, as well as providing convincing evidence of correctness. The case study also demonstrates the problems of using formal proof in this process and recommends ways in which proofs may be managed to retain their quality as convincing evidence without leading to excessive detail.

## 1 Introduction

There are two reasons for carrying out the process of formal refinement, one is to help in the design of software and the other is to provide evidence of correctness to an independent evaluator. While the concept of proof of correctness is the basis for the formal method of refinement, a different emphasis will be given to proof according to thereason for using formal refinement. For the designer, formal proofs are of little help. Indeed, because the production of a formal proof is a difficult and unpredictable process it actually impedes the production of software. On the other hand, proofs are very important to evaluators as, in principle, they provide convincing evidence of correctness and so it is accepted that for high integrity software formal proofs will be necessary. However, even here one can question the extent to which the formalisation helps in demonstrating trustworthiness. In principle a formal proof forms a convincing argument but the details of the formalism may obscure the understanding of what is being proved. By hiding detail and automatically checking proof steps, tools could assist the process but with current technology the degree to which a proof helps rather than hinders the evaluation is an open question.

This does not mean that the formal method of refinement is valueless. For both design and evaluation, refinement provides the concepts and mathematical basis necessary for intellectual grasp of the software. Assurance is based on understanding and the lessons learnt from the case study described in this paper are that the practice of refinement should support this. Because we *are* concerned with practice, the case study needs to deal with a non-trivial problem, which is concerned with pattern matching in the programming language ML. A complete description of the formalisation is contained in [Macdonald *et al* 1989] which includes an implementation of the problem in Algol 68. In this paper, details of the formalisation will be omitted where the development is straightforward.

Section 2 presents a formal specification of pattern matching in ML. This is a problem which is interesting in its own right, but the formal refinement has a number of surprises which demonstrate the power of the method over the use of an intuitive approach.

Sections 3 to 5 deal with the formal development of the design specification. Section 3

---

[*]Royal Signals and Radar Establishment, Malvern, Worcestershire, WR14 3PS, United Kingdom
Copyright © Controller HMSO London 1990

deals with the choice of representation for the implementation. For the refinement method, the relationship between the the entities in the abstract specification and those in the concrete implementation must be given and this is known as the abstraction invariant. Given this, the constraints on the implementation functions can actually be calculated, and this is done in section 4. With these in mind, section 5 deals with the design of the implementation algorithms at an abstract level which ensures that they are compatible with the specification. It is this aspect of abstract algorithm design where refinement has most to offer the implementor as opposed to the evaluator.

Section 6 covers the evaluator's requirements. It discusses the type of proof document which would illuminate the discussion rather than obscuring it and gives an example of the way in which the salient features of the proof requirements could be displayed.

Finally section 7 discusses the advantages and disadvantages of the formal approach and gives recommendations for current practice.

As some readers may not be familiar with ML the remainder of this introduction describes the pattern matching features of the language. ML was produced as a spin-off from the theorem proving system LCF [Gordon et al 1979, Paulson 1987]. In the latter, theorems, proof rules, tactics and so on are manipulated by means of an interactive *meta language*. As the ideas for this crystallised it was realised that the best meta language was in fact a programming language enriched with a polymorphic type discipline and exception handling. The language could be used quite generally in a similar manner to Lisp. The LCF idea produced a number of offspring, and varying dialects of ML began to appear. As a result, Milner convened the ML community with a view to developing a standard [Harper *et al* 1988]. The new language is larger and more powerful than its predecessors and is interesting in its own right, quite apart from its exciting use with theorem proving.

An attractive feature of the language is the use of *pattern matching* in the specification of functions. In ML, a function may be supplied as a series of clauses, each clause specifying how values of a certain structure are to be handled by the function as a whole. This is best illustrated by means of an example, for which the ML concepts of datatypes, patterns and pattern matching will be needed.

The ML *datatype* declaration is similar to a disjoint union and is a generalisation of the idea of an enumerated type in Pascal. The simplest datatype declarations take the form

```
datatype colour = red | blue | green
```

which declares *colour* as a type containing the three values red, blue and green. More usually, the datatype is constructed from previously defined types, as in the following declaration:

```
datatype label = code of colour | number of int*int | name of string
```

Thus a label value is constructed either from a colour, a pair of integers or a string. Instead of being constants as in the previous example, code, number and name are *constructor functions* which construct values having the ML type label. So code(red), number(1, 2) and name "fred" are all values of this type.

An ML pattern is either a variable, an expression involving constants, constructor functions and variables or is a tuple of patterns. Thus, given variables x, i, j, then x, number(i, j) and code(red) are all patterns. A pattern *matches* a set of values. A variable will match any value, but tuple patterns and patterns involving constructors only match values which have the corresponding structure. Thus x matches any value, number(i, j) matches any label value constructed from two integers and code(red) matches the single value it denotes.

A simple example of a function declared as a series of clauses is one which permutes the colours:

```
fun   colour_perm red = blue
   |  colour_perm blue = green
   |  colour_perm green = red
```

This is both clearer and more concise than the corresponding expression involving conditionals. The language specification requires an ML compiler to check that the patterns supplied in the parameter position account for all the possible values to which the function might be applied. Thus `colour_perm` is a function from `colour` to `colour`, but if the last clause of its definition had been omitted it would have been applicable to `red` and `blue` values only. Consequently ML compilers must be able to check whether a set of patterns is *exhaustive* in this way. A related problem is that the addition of a clause to a function may be *redundant*, that is, it may not increase the number of values already matched. In this case the clause is superfluous and the programmer has made a mistake.

The two tests required are intuitively obvious in the simple case above, but when constructor functions and tuple values are considered, intuition can be misleading. In this case the formal specification is particularly helpful and it is presented here using the specification language Z [Sufrin 1983, Hayes 1987, Spivey 1988]. The particular syntax used is that specified in [Sennett 1987] which is developed from that described in [King 1987].

## 2 Z specification of the problem

### 2.1 specification of ML values and types

The specification is defined in terms of the operations a compiler must perform to carry out the tests required. Three operations are defined, an initial operation used to specify the compiler's action on encountering the first pattern of the clausal function, a testing operation to specify the action required on encountering the subsequent patterns and a final operation to test whether the set of patterns is exhaustive. The parameters for the second operation are the new pattern just compiled and the set of values matched by the patterns obtained from the other clauses of the function (the other patterns in the *match*, in ML terms). The check to be made is whether the set of values matched increased as a result of adding the latest pattern, in which case the match is not redundant. For the third operation, the parameter is the set of values accounted for by all the patterns in the match, and the test is whether this is all the values belonging to the type, in which case the set of patterns is exhaustive.

The problem is defined in terms of ML values so it is necessary to build a Z representation of them. In this a number of significant simplifications may be made. First of all, the treatment of the special constants (denotations for integers, reals and strings) is the same for each type of constant, so they are all represented together by the Z given set *SCONST*. Secondly, function values may only be matched by variables, which by definition are exhaustive. As the test is trivial in this case, function values are not considered further (they may be considered as being members of *SCONST*). Thirdly, the compiler's representation of constructors is irrelevant to the specification of the problem, so the constructors are simply introduced as a given set *CON*. Fourthly, the ML record type will simply be treated as though it were the corresponding tuple. Finally, polytypes will be ignored. A polymorphic match will be exhaustive or redundant at each and every instance of its type, so the problem is independent of the polymorphism.

With these simplifications, an ML value is either the special ML void or unit value or is drawn from the set of special constants or is a construction or a tuple of values. For the Z representation, a labelled, disjoint union will be used, but as the definition is recursive, *Value* is introduced as a given set. Thus values will be built up from the following given sets:

[*Value, SCONST, CON*]

An exhaustive set of patterns is one which matches every value of a given type and so the next step is to define what is meant by a type. The types are introduced as disjoint subsets of the values, with the actual type structure being given by constraints to be added later:

$Type : \mathbb{P}\,\mathbb{P}_1\,Value$

$\bigcup Type = Value$
$\forall T_1, T_2 : Type \mid T_1 \neq T_2 \bullet T_1 \cap T_2 = \emptyset$

A constructed value is built from a constructor and a value for the parameter. The parameter value must be drawn from a specific type which was associated with the constructor in the datatype declaration. This association can be represented by the following (unspecified) function:

$type\_of\_con : CON \rightarrow Type$

Constructed values may now be described in terms of the following schema:

$$\begin{array}{|l}\hline \textit{Cons\_val} \\ \hline con : CON;\ val : Value \\ \hline val \in type\_of\_con(con) \\ \hline \end{array}$$

A constructor which is a constant is treated as having the unit value as a parameter.

Tuples are represented as sequences of two or more values. Tuples of patterns will also be required so it is helpful to make the generic definition:

$$\begin{array}{|l}\hline [T] \\ \hline tuple\ T == \{s : seq\ T \mid \#s > 1\} \\ \hline \end{array}$$

With these definitions, the structure of values may now be represented by

$Value ::= unitv \mid sconstv \langle\!\langle SCONST \rangle\!\rangle \mid consv \langle\!\langle Cons\_val \rangle\!\rangle \mid tupv \langle\!\langle tuple\ Value \rangle\!\rangle$

Given this simplified model of the values, the type structure is determined only by the constructors and tuples, and members of the same type will have the same constructor and tuple structure. This property may be specified as a relation *same_type* on the values.

$\_same\_type\_ : Value \leftrightarrow Value$

This may be defined straightforwardly by cases according to the structure of *Value*.

The set of types is simply the set of equivalence classes of *same_type*:

$Type = \{ty : \mathbb{P}\ Value$
$\quad \mid \forall v_1 : ty;\ v_2 : Value$
$\quad \bullet\ v_2 \in ty \land v_1\ same\_type\ v_2 \lor v_2 \notin ty \land \neg v_1\ same\_type\ v_2$
$\quad \}$

Consequently, it is possible to deduce the type of a value:

$type\_of == \lambda\ v : Value \bullet \{w : Value \mid w\ same\_type\ v\}$

## 2.2 Specification of ML patterns and pattern matching

Patterns are formed in a similar way to values, except that one of the primitive patterns is a variable which matches any value. As with values, simplifications will be made: the ML layered pattern feature will be ignored and the wild card treated as a variable. For this problem it is not necessary to know the representation of variables, so they are introduced as a given set along with *Pattern* itself, introduced for the purposes of the recursive definition:

[*Pattern, Variable*]

Constructed patterns may be formed using the following schema:

$$\begin{array}{|l}\hline \textit{Cons\_patt} \\ \hline con : CON;\ patt : Pattern \\ \hline \end{array}$$

Using this, the structure of patterns is given by

*Pattern ::= unitp | sconstp « SCONST » | var « Variable » | consp « Cons_patt »*
*| tupp « tuple Pattern »*

The definition of pattern matching is given in terms of a relation, *matches*, specifying which values match a given pattern:

_matches_ : *Pattern* ↔ *Value*

Again, this may be defined recursively over the structure of patterns and values in a straightforward fashion. For the special constants a simplification was made. The set of special constants is infinite and can never be matched by a finite set of patterns unless it contains a variable. For this case, it seems excessive to keep track of the special constants in a set of patterns simply in order to check for redundancy. Consequently the specification is relaxed to omit the redundancy check in this case. This was done by allowing special constants to be matched by variables only. (Note that this is not necessary to our approach, but apart from being a sensible relaxation it simplifies the presentation.)

The value coverage of a set of patterns is the set of values which may be matched to the patterns. As a variable matches any value it is necessary to restrict the set of values to one type so the compiling operations are specified in terms of a *valcover* function as follows:

*valcover* == λ *patt* : *Pattern*; *ty* : *Type* • {*v* : *ty* | *patt matches v*}

For a set of patterns *patts* of type *ty* to be exhaustive

∪ {*p* : *patts* • *valcover(p, ty)*} = *ty*

## 2.3 The compiling operations

The implementation of the compiling operations is considerably simplified if it is possible to make use of the fact that the patterns are well-typed. This constraint may be expressed using a relation, similar to *same_type*, which specifies which patterns are compatible with a type. This relation may be defined recursively in the usual way and has a type given by:

_pattern_compatible_ : *Pattern* ↔ *Type*

For convenience, the predicate is defined as a schema:

```
┌─ Pattern_Compatible ─────┐
│ p : Pattern; type : Type │
├──────────────────────────┤
│ p pattern_compatible type│
└──────────────────────────┘
```

The initial compiling operation generates the set of values from the first pattern to be compiled:

```
┌─ Init_op ──────────────────────────────────────────────┐
│ vals! : ℙ Value; par? : Pattern; type : Type           │
├────────────────────────────────────────────────────────┤
│ par? pattern_compatible type ∧ vals! = valcover(par?, type) │
└────────────────────────────────────────────────────────┘
```

The type of the pattern being compiled is given by *type*, and as we are not concerned with type checking aspects of the compiler, this is treated as a constant throughout all the pattern checking operations.

It is convenient to express the result of the second compiling operation in terms of a new Z datatype:

*Result* ::= *OK* | *INCOMPLETE* | *REDUNDANT*

The operation is specified in terms of the set of values covered by the patterns compiled so far and the effect of adding one more pattern.

─ *Check_op* ────────────────────────────────
*vals?, vals!* : $\mathbb{P}$ *Value*
*par?* : *Pattern*
*r!* : *Result*
*type* : *Type*
────────────────────────────────
*par? pattern_compatible type* $\wedge$ *vals?* $\subseteq$ *type*
*vals!* = *valcover(par?, type)* $\cup$ *vals?*
*valcover(par?, type)* $\neq$ {} $\wedge$ *vals!* = *vals?* $\wedge$ *r!* = *REDUNDANT*
$\vee$
(*valcover(par?, type)* = {} $\vee$ *vals!* $\neq$ *vals?*) $\wedge$ *r!* = *OK*
────────────────────────────────

In this schema, *vals?* represents the set of values accounted for by the patterns compiled so far and *vals!* the result of adding the new pattern *par?*. The result of the operation is left in *r!*. The predicate *valcover(par?, type)* $\neq$ {} eliminates reporting a redundancy when the patterns include special constants.

The pre-condition of this operation is easily simplified to

*par? pattern_compatible type* $\wedge$ *vals?* $\subseteq$ *type*

If this is so, then from the definition of *valcover* it follows that *vals!* is also a subset of *type*, and for both operations. This is essential as the placing of the compiling operation with respect to the syntax of ML ensures that the output of *Init_op* will form the input to *Check_op* as does the output of *Check_op* itself. Strictly speaking, *type* is redundant in *Check_op* as it could be deduced from *vals?*. (A function *type_of*, which gives the type of a value, has already been defined.) However, leaving the type in in this way leads to a clearer specification.

The final operation checks whether the set of patterns is exhaustive and is simply given by:

─ *Final_op* ────────────────────────────────
*vals?* : $\mathbb{P}$ *Value*;  *type* : *Type*;  *r!* : *Result*
────────────────────────────────
*vals?* = *type* $\wedge$ *r!* = *OK* $\vee$ *vals?* $\neq$ *type* $\wedge$ *r!* = *INCOMPLETE*
────────────────────────────────

As with the other operations, the ML syntax determines when this operation is called: the input is provided either by *Init_op* for a one-pattern match, or by the last call of *Check_op* in a multi-pattern match.

## 3 Formal implementation - the abstraction invariant

The specification has been written in a form which matches as closely as possible the informal specification given in the language definition. As such it is defined in terms of possibly infinite sets of values and it is not possible to implement a test such as *vals?* = *type* in *Final_op* directly. Instead a coverage measure is used to keep track of how completely a set of patterns accounts for the values in a type. Data refinement is used to specify the operations on coverages which correspond to the operations in the specification and for this an abstraction invariant is defined which specifies the set of values which correspond to a given coverage.

The first question to be settled is the form for the coverage measure. It must be easily implementable and the check not excessively time-consuming. The essentials of the problem are that for constructed patterns every constructor must be accounted for and for tuple patterns the individual elements must be complete. This latter property is surprisingly hard to formalise, so it is useful to have a few test cases to clarify what is actually required. Using the datatype definitions given previously, consider the case of a 3-tuple of colours. The following sequence of patterns is complete and not redundant in the sense that each succeeding pattern matches more values and the complete set of values is only accounted for with the final pattern.

|  | Number of extra values matched |
|---|---|
| (red, blue, green) | 1 |
| (x, blue, green) | 2 |
| (red, x, green) | 2 |
| (red, blue, x) | 2 |
| (red, x, y) | 4 |
| (blue, green, red) | 1 |
| (blue, x, y) | 7 |
| (x, green, green) | 1 |
| (x, red, green) | 1 |
| (x, y, z) | 6 |

It is worth while checking this table to convince yourself that if the concept of exhaustive patterns is intuitive, the actual check to apply in the tuple case is not. Tuples present problems because the combination of the values covered by one element of the tuple pattern with those covered by another is difficult to grasp in the general case of n-tuples. Accordingly the coverage measure for tuples uses a function from the coverage provided by the first element of the tuple to the coverage provided by the remainder of the tuple.

The other complication in the problem is that it is necessary to account for constructor functions as well as constants in the patterns. Thus for a pair of labels one has the following sequence of exhaustive but not redundant patterns:

```
(code(red), x)
(x, number(0, y))
(code(x), y)
(x, name(y))
(x, y)
```

It is possible to treat constructed patterns as though they were 2-tuples, using a function in exactly the same way as for tuples. This approach has not been taken in the implementation presented here, partly for reasons of efficiency and partly for ease of understanding. Instead, a function from constructors to coverages is used.

Special constants are treated by having an *incomplete* coverage measure, which represents an empty set of values, corresponding to the fact that, in the specification, a special constant pattern matches no value. Treating the special constants accurately would require keeping a measure dependent on the number of constants already accounted for.

With this motivation the Z datatype defining the coverage measure may be written down:

*Cover ::= complete | incomplete*
    *| construct* ⟪ *CON* ↛ *Cover* ⟫ *| pair* ⟪ $\mathbb{F}$ *(Cover × Cover)* ⟫

Thus the coverage measure chosen is a tree in which the leaves correspond to full or no coverage and the nodes of the tree correspond to the constructor and tuple structure of the type. The parameter of *pair* is given as a set of pairs rather than a function as this seems to make the explanation easier. Finite sets are used to ensure that the datatype is satisfiable.

The formal definition of the abstraction invariant will be motivated by giving an example of the intended relation between coverages and patterns. The first two patterns from the sequence of label pairs above are to be represented by the following two coverages:

*pair* {*construct* {*code* ↦ *construct* {*red* ↦ *complete*}} ↦ *complete*}
*pair* {*complete* ↦ *construct* {*number* ↦ *pair* {*incomplete* ↦ *complete*}}}

The essence of the problem is to combine successive coverages into one in such a way as to end up with the *complete* cover when the set of patterns is exhaustive.

The basic requirement for the abstraction is a function from a *Cover* to a set of *Value*. This cannot be provided from the datatype chosen because the set of values from a *complete* cover will depend on the type. However, the actual check to be made is independent of the type in this case, so the specification contains redundant information as far as these particular checks are concerned. (This state of affairs is called *bias*.) It is still possible to carry out the refinement, but in order to do so, it is necessary to have a function from a *Cover* and a *Type* to a set of *Value*. A unique definition of this function will be provided, built up according to the structure of *Cover* and *Type* in the usual way.

$Abs\_fn : (Cover \times Type) \nrightarrow \mathbb{P}\ Value$

For the primitive coverage elements we have the following

─ *AIPrimitive* ─────────────────────────────
$c : Cover;\ type : Type;\ vals : \mathbb{P}\ Value$
───────────────────────────────────────────
$(c = complete \wedge vals = type) \vee (c = incomplete \wedge vals = \{\})$
───────────────────────────────────────────

For constructed coverages the cover represents any constructed value it is possible to form from a constructor in the domain of the constructor function combined with a value drawn from the set represented by the corresponding coverage element in the range of the function. This is expressed by the following schema:

─ *AIConstructions* ─────────────────────────
$c : Cover;\ type : Type;\ vals : \mathbb{P}\ Value$
───────────────────────────────────────────
$c \in ran\ construct \wedge type \subseteq ran\ consv$
$vals = \{Cons\_val;\ F : CON \nrightarrow Cover$
    $|\ c = construct\ F \wedge con \in dom\ F$
    $\wedge\ val \in Abs\_fn(F\ con, type\_of\_con(con))$
    $\bullet\ consv\ \theta Cons\_val$
    $\}$
───────────────────────────────────────────

For tuples some functions for manipulating tuple types are necessary. From a tuple type one can form a type corresponding to the first element of the tuple and one corresponding to the remainder of the tuple: the functions carrying out this operation are called *HD* and *TL* respectively. It is also necessary to have a function from pairs of coverages to sequences of values as follows:

$$\text{Tuple\_vals} : (\text{Cover} \times \text{Cover} \times \text{Type}) \twoheadrightarrow \mathbb{P} \text{ tuple Value}$$

$\forall c_1, c_2 : \text{Cover}; \text{ type} : \text{Type}; \text{ vals} : \mathbb{P} \text{ tuple Value}$
$| \text{ type} \subseteq \text{ran tupv}$
$\bullet \text{ Tuple\_vals}(c_1, c_2, \text{type}) = \text{vals}$
$\Leftrightarrow$
$\text{Size type} = 2 \land \text{vals} = \{v_1 : \text{Abs\_fn}(c_1, \text{HD type}); v_2 : \text{Abs\_fn}(c_2, \text{TL type})$
$\bullet \langle v_1, v_2 \rangle$
$\}$
$\lor$
$\text{Size type} > 2 \land \text{vals} = \{v_1 : \text{Abs\_fn}(c_1, \text{HD type}); pc_1, pc_2 : \text{Cover};$
$tv : \text{tuple Value}$
$| pc_1 \mapsto pc_2 \in \text{pair}^{-1} c_2$
$\land tv \in \text{Tuple\_vals}(pc_1, pc_2, \text{TL type})$
$\bullet v_1 \text{ cons } tv$
$\}$

The set of values corresponding to a pair of covers is that set of tuple values formed by taking any element from the set corresponding to the first cover for the first element and combining it with any element from the set corresponding to the second. Thus if the first cover corresponds to $n$ values and the second to $m$, the pair of covers corresponds to $n \times m$ tuple values. For longer tuples, the second cover value will itself be a pair giving rise to a set of tuple values whose size is one less than the original. Form a set by taking any element from the values corresponding to the first cover and add this to the front of any element in the tuples provided by the second cover.

With this auxiliary function the abstraction function for tuples may be defined as follows:

___ AITuples ___
$c : \text{Cover}; \text{ type} : \text{Type}; \text{ vals} : \mathbb{P} \text{ Value}$

$c \in \text{ran pair} \land \text{type} \subset \text{ran tupv}$
$\text{vals} = \bigcup \{c_1, c_2 : \text{Cover} | c_1 \mapsto c_2 \in \text{pair}^{-1} c \bullet \text{tupv} \llbracket \text{ Tuple\_vals}(c_1, c_2, \text{type}) \rrbracket \}$

The abstraction function itself is simply given by the constraint:

$\text{Abs\_fn} = \lambda c : \text{Cover}; \text{ type} : \text{Type}$
$\bullet \mu \text{ vals} : \mathbb{P} \text{ Value} | \text{AIPrimitive} \lor \text{AIConstructions} \lor \text{AITuples} \bullet \text{vals}$

Finally, it is convenient to define the abstraction invariant as a schema:

___ AI ___
$c : \text{Cover}; \text{ type} : \text{Type}; \text{ vals} : \mathbb{P} \text{ Value}$

$\text{vals} = \text{Abs\_fn}(c, \text{type})$

# 4 The refined operations

## 4.1 The initial operation

It is interesting to follow the technique given in Morgan [1988] in which the concrete operations corresponding to the abstract operations are actually calculated from the abstraction invariant. To summarise the notation, which will be slightly adapted from that used by Morgan, an operation is represented by its pre and post-conditions as below:

$$\Delta\ av\ [pre, post]$$

This represents an operation achieving a state of affairs specified by the predicate *post*. It must be given an initial state represented by *pre* and achieves it by altering the abstract variable *av*.

Given an abstraction invariant *AI*, involving the concrete variable *cv*, the corresponding concrete operation is simply given by the formula

$$\Delta\ cv\ [\exists\ av \bullet AI \wedge pre,\ \exists\ av \bullet AI \wedge post]$$

This technique will be applied first of all to the operation *Init_op*. In the refinement notation we can write:

$$Init\_op \sqsubseteq \Delta\ vals\ [par\ pattern\_compatible\ type,\ vals = valcover(par, type)]$$

This statement is an assertion that the operation on the right hand side of the ⊑ symbol is an operation refinement of the Z schema operation *Init_op*. In the refinement notation, operations are expressed in terms of variables assumed to be declared within the current context of the operation. Variables in the pre-condition refer to values before the operation while variables in the post-condition refer to values after. Consequently, there is no need for the Z decorations of !, ? or ' and these are systematically dropped. The pre-condition for the operation has been derived by existentially quantifying over the output variables in the Z schema and simplifying.

For the data refinement, the concrete variable is the coverage, *c*, of type *Cover* while the abstract variable is the set of values *vals*, which has the type $\mathbb{P}\ Value$. The concrete operations will eventually prove to be independent of the type which may simply be discarded. Using the abstraction invariant and the data refinement symbol ⪯, our operation is refined as follows:

$$\preceq \Delta\ c\ [\exists\ vals : \mathbb{P}\ Value \bullet AI \wedge par\ pattern\_compatible\ type,$$
$$\exists\ vals : \mathbb{P}\ Value \bullet AI \wedge vals = valcover(par, type)]$$

This operation is guaranteed to be a correct data refinement of *Init_op*. Looking at the post-condition, it is clear that it is necessary to calculate a coverage from the input pattern. This coverage must be such that the application of the abstraction function gives the same set of values as that provided by *valcover* when applied to the input pattern. It is fairly obvious, once one has stood back from the trees of the formalism to view the wood of the problem, that the initial value of *c* is irrelevant. Consequently there is no problem incurred in weakening the pre-condition and simplifying the post-condition by substituting for *vals* as follows

$$\sqsubseteq \Delta\ c\ [par\ pattern\_compatible\ type,\ Abs\_fn(c, type) = valcover(par, type)]$$

The implementation problem therefore is to define a *coverage* function which relates a *Pattern* to a *Cover* in the manner required:

$$coverage : Pattern \rightarrow Cover$$

For this to be a correct implementation of the initial operation, the following theorem has to be proved:

*Pattern_Compatible*
⊢
*Abs_fn(coverage(p), type) = valcover(p, type)*

The specification of *coverage* will be deferred to a later section because the other operations introduce further constraints on its definition.

## 4.2 The checking operation

Proceeding in the same way as before

$$Check\_op \;\hat{=}\; con\; vals_0 \bullet \Delta\; vals,\, r\; [vals = vals_0 \wedge par\; pattern\_compatible\; type \\ \wedge\; vals \subseteq type,\; Check\_op]$$

This step has introduced some more refinement notation. Where, as in this case, the post-condition is dependent on the values both before and after the operation, it is necessary to preserve the initial value using a logical constant, which is introduced with the reserved word con. By convention, initial values are indicated with a 0-subscript, so $vals_0$ corresponds to the input value *vals?* in the schema.

Note that in this specification the pre-condition does not record the fact that the input values are provided from the results of the initial operation or a previous use of the checking operation as the case may be. This will be introduced informally later. The data refinement for the checking operation can be written as:

$$\preceq con\; vals_0,\, c_0 \bullet \Delta\; c,\, r\; [\exists\; vals : \mathbb{P}\; Value \\ \bullet\; AI \wedge vals = vals_0 \wedge c = c_0 \\ \wedge\; par\; pattern\_compatible\; type \wedge vals \subseteq type, \\ \exists\; vals : \mathbb{P}\; Value \bullet AI \wedge Check\_op]$$

Because *Abs_fn(c, type)* is always a subset of *type*, the pre-condition may be replaced by $vals_0 = Abs\_fn(c_0, type) \wedge par\; pattern\_compatible\; type$.

The post-condition may be simplified, by eliminating *vals* and $vals_0$, into a predicate described by the following schema:

―― *Check_op_1* ――――――――――――――――――――――
$c, c_0 : Cover$
$par : Pattern$
$type : Type$
$r : Result$
――――――――――――――――――――――――――
*par pattern_compatible type*
$Abs\_fn(c, type) = valcover(par, type) \cup Abs\_fn(c_0, type)$
  $valcover(par, type) \neq \{\} \wedge Abs\_fn(c, type) = Abs\_fn(c_0, type)$
  $\wedge\; r = REDUNDANT$
$\vee\; (valcover(par, type) = \{\} \vee Abs\_fn(c, type) \neq Abs\_fn(c_0, type))$
  $\wedge\; r = OK$
――――――――――――――――――――――――――

We already have the requirement to find a coverage function such that *valcover(par?, type) = Abs_fn(coverage(par?), type)*, so it is tempting to define a union function between coverages which carries out the corresponding operation to forming a union of sets of values:

$union : (Cover \times Cover) \nrightarrow Cover$

and refine to the operation

```
┌─ Check_op_2 ─────────────────────────────────────┐
│ c, c₀ : Cover                                     │
│ par : Pattern                                     │
│ type : Type                                       │
│ r : Result                                        │
├───────────────────────────────────────────────────┤
│ c = union(coverage(par), c₀)                      │
│   coverage(par) ≠ incomplete ∧ c = c₀ ∧ r = REDUNDANT │
│ ∨ (coverage(par) = incomplete ∨ c ≠ c₀) ∧ r = OK  │
└───────────────────────────────────────────────────┘
```

For this to be so, the following theorem must be proved:

Check_op_2
⊢
Check_op_1

This corresponds to strengthening the post-condition, namely, that the achievement of *Check_op_2* will entail the achievement of *Check_op_1*. Note that the type has now dropped out of the predicates.

For the proof of this theorem, it is necessary to show that the *union* function behaves like set union and that there is a unique representation of the empty set of values. It is difficult to define a function having the properties required without taking into account the fact that the coverages have all been derived from patterns of the same type. Consequently, extra constraints will be added to the specification which are satisfied by the pre-condition. These constraints will be defined in terms of a *cover_compatible* relation, and express the fact that the *union* function is only required to combine coverages of the same structure. This may defined in a similar way to *pattern_compatible* and has type:

_cover_compatible_ : Cover ↔ Type

In the refined operation, the compatibility of the coverage $c$ with the type is guaranteed by the compatibility of the pattern *par* with the type while the fact that $c_0$ is compatible is guaranteed by the initial set of values in the abstract operation being generated by matching type-compatible patterns. This is all rather tedious to formalise and not very illuminating, so these theorems will not be stated. The theorems expressing the more interesting union and uniqueness properties may be broken down into sub-goals defined in terms of the following schema, which gathers together the parameters of *union* and its result, and has the type as a parameter:

```
┌─ Union ──────────────────┐
│ c₁, c₂, c : Cover         │
│ type : Type               │
├──────────────────────────┤
│ c = union(c₁, c₂)         │
│ c₁ cover_compatible type  │
│ c₂ cover_compatible type  │
└──────────────────────────┘
```

The first goal is related to the first constraint of *Check_op_2*:

*Union*
⊢
$Abs\_fn(c, type) = Abs\_fn(c_1, type) \cup Abs\_fn(c_2, type)$

The remaining goals are related to the second constraint:

*Pattern_Compatible*
⊢
$coverage(p) \neq incomplete \Rightarrow valcover(p, type) \neq \{\}$

*Union*
⊢
$c = c_2 \Rightarrow Abs\_fn(c, type) = Abs\_fn(c_2, type)$

*Union*
⊢
$c \neq c_2 \Rightarrow Abs\_fn(c, type) \neq Abs\_fn(c_2, type)$

The first of these goals requires the *coverage* function to deliver the *incomplete* value whenever the value coverage is empty. The second is trivially true and the third requires the result of the *union* function to change whenever the set of values covered changes. Further constraints on *union* will emerge with the definition of the final operation.

## 4.3 The final operation

This performs the final check for whether the set of patterns is exhaustive or not. The operation is simply refined as follows:

$Final\_op \sqsubseteq \Delta\, r\, [true, vals = type \land r = OK \lor vals \neq type \land r = INCOMPLETE]$
$\preceq \Delta\, r\, [true, c = complete \land r = OK \lor c \neq complete \land r = INCOMPLETE]$

Here the data refinement and simplification of the pre and post conditions have been carried out in one step. The input to the operation is simply the result of *union*, or, in the case of a single clause in the function definition, the result of *coverage*. Accordingly, there are further constraints to satisfy. These are:

*Union*
⊢
$c = complete \Rightarrow Abs\_fn(c, type) = type$

*Union*
⊢
$c \neq complete \Rightarrow Abs\_fn(c, type) \neq type$

$p : Pattern;\ type : Type;\ c : Cover$
⊢
$c = coverage(p) = complete \Rightarrow Abs\_fn(c, type) = type$

$p : Pattern;\ type : Type;\ c : Cover$
⊢
$c = coverage(p) \neq complete \Rightarrow Abs\_fn(c, type) \neq type$

The first and third of these goals follow immediately from the definition of the abstraction function, while the second and fourth again require a unique representation of completeness.

# 5 Abstract algorithm design

So far, apart from the definition of the coverage measure, the steps taken have been mechanical and the advantage of the formalisation has been with the precision it offers. For the creative design steps, the formal method also provides the concepts which need to be implemented and hence makes a major contribution to the rational design process. Apart from the need to test for complete and incomplete partial results, the coverage function is fairly straightforward, so we shall concentrate on the *union* function. For this, it is required to provide a constructive definition of a function which satisfies the following theorems:

*Union*
⊢
*Abs_fn(c, type)* = *Abs_fn($c_1$, type)* ∪ *Abs_fn($c_2$, type)*

*Union*
⊢
$c = c_2$ ⇒ *Abs_fn(c, type)* = *Abs_fn($c_2$, type)*

*Union*
⊢
$c \neq c_2$ ⇒ *Abs_fn(c, type)* ≠ *Abs_fn($c_2$, type)*

*Union*
⊢
*c = complete* ⇒ *Abs_fn(c, type)* = *type*

*Union*
⊢
*c ≠ complete* ⇒ *Abs_fn(c, type)* ≠ *type*

The principal objective is to make the *union* function correspond to the operation of uniting sets of values. The additional constraints are that the coverage must only change when the underlying sets of values change and that it is necessary to have a unique measure for the exhaustive set of patterns.

For the abstract algorithm design, consider first the case of a set of constructed patterns. These are represented by a *construct* coverage which measures the coverage of values associated with each of the constructors in the datatype. So the coverage will be represented by some function, $F$, of the form $F = \{c_i \mapsto P_i\}$, where the $c_i$ are the constructors of a datatype and the $P_i$ represent sets of parameter values covered so far. Adding a new pattern will give rise to an extra coverage represented in the same way by the maplet $c_j \mapsto P$. The new coverage, $F'$, will depend on whether $c_j$ is a member of the domain of $F$ or not. If it is, the coverage provided by the parameter of the new pattern is united with the coverage provided by the parameters of the patterns already processed which use that constructor. This is expressed formally as $F' = F \oplus \{c_j \mapsto P_j \cup P\}$. When a new pattern introduces a constructor for the first time, the coverage is simply added to those already there. In this case, $F' = F \cup \{c_j \mapsto P\}$.

Calculating the united coverage in this way will cause $F'$ to differ from $F$ exactly when a new constructor is added to the coverage or when a parameter coverage, *parc*, changes. This satisfies the requirement of the third goal for this case. For the fifth goal, a test for completeness is required, expressed as follows:

⎡ *Constructor_complete* ───────────────
│ $F' : CON \nrightarrow Cover$
│────────────────────────────────────
│ ∃ $cs$ : ran *datatype* • $F' = \{c : cs \bullet c \mapsto complete\}$
⎣────────────────────────────────────

Using this schema and generalising to cover the case of merging sets of patterns, the *union* constructor case is defined as follows:

―― UConstructions ――――――――――――――――――
$c_1, c_2, c :$ Cover
――――――――――――――――――――――――――――
$c_1 \in$ ran construct $\land\ c_2 \in$ ran construct
Constructor_complete $\land\ c =$ complete
$\lor$
$\neg$Constructor_complete $\land\ c =$ construct $F'$
where
$\quad F == \text{construct}^{-1} c_1$
$\quad f == \text{construct}^{-1} c_2$
$\quad f_1 == \text{dom } f \mathbin{⩤} F$
$\quad f_2 == \text{dom } F \mathbin{⩤} f$
$\quad f_3 == \{c : \text{dom } F \cap \text{dom } f \bullet c \mapsto \text{union}(F\ c, f\ c)\}$
$\quad F' == (f_1 \cup f_2 \cup f_3)$
――――――――――――――――――――――――――――

This is fairly straightforward, but the same technique may be used to deal with tuple patterns, which are represented by *pair* coverages. In this case, one coverage stands for the set of values covered in the first element of the tuple, which is bound to the set of values covered by the rest of the tuple in a similar way to that in which the parameter of a constructed pattern is bound to its constructor. Consequently a pair coverage stands for a set of values which may be represented in the form $F = \{A_i \mapsto B_i\}$, where the $A$s and $B$s now stand for sets of values. Each element of the set $F$ stands for a set of tuple values formed by taking one element out of an $A$ and combining it with any element out of the corresponding $B$. A new pattern gives rise to a set of values covered of the form $\alpha \mapsto \beta$. For a given element of $F$, say $A \mapsto B$, the standard rules for taking unions of Cartesian products should give rise to the following extra elements in $F'$, for the addition of this set:

$\quad A \setminus \alpha \mapsto B$
$\quad A \cap \alpha \mapsto B \cup \beta$
$\quad \alpha \setminus A \mapsto \beta$

Applying this procedure to every element in $F$ gives the new coverage, $F'$. If any of the sets are empty, this element represents no values and may be discarded. To meet the other goals it is important that $F'$ should differ from $F$ exactly when new values have been added to the set of values covered. This objective is attained by keeping the $A_i$ disjoint: an alteration to one of them cannot then produce a value which is already accounted for by the other members of $F$. If this is the case, and if the first two operations above correspond to the introduction of new values, that is, if $B \cup \beta \neq B$, alterations must correspond to new values. If all the $A_i$ in $F$ are disjoint, elements formed by these operations will also be disjoint. However, the third operation may give rise to intersections, but these may be eliminated by instead adding one element $(\alpha \setminus \bigcup (\text{dom } F)) \mapsto \beta$ to the function as a whole, rather than carrying out the operation for each element.

As an example of this process, consider the following sequence of patterns:

```
(red, y)
(green, blue)
(x, red)
(green, green)
(blue, y)
```

These patterns match the following pairs of values:

$$red \mapsto colour$$
$$green \mapsto blue$$
$$colour \mapsto red$$
$$green \mapsto green$$
$$blue \mapsto colour$$

Uniting the pairs according to these rules gives successively:

$$\{red \mapsto colour, green \mapsto blue\}$$
$$\{red \mapsto colour, green \mapsto \{red, blue\}, blue \mapsto red\}$$
$$\{\{red, green\} \mapsto colour, blue \mapsto red\}$$
$$colour \mapsto colour$$

In formalising this process it is necessary to specify operations representing the difference and intersection of sets of values, just as it is already required to form unions. For differences and intersections, it is necessary to distinguish the empty set, which corresponds to the *incomplete* cover. The difference and intersection functions are both defined recursively and their types are given by:

*difference, intersection* : *(Cover × Cover)* $\nrightarrow$ *Cover*

Difference and intersection of tuples involves sets of pairs of differences, so it useful to define some schemas to provide the necessary signatures as follows:

```
┌─ Pair_element ────────────────┐
│ A₁, A₂, α₁, α₂ : Cover         │
│ res_pairs : ℙ (Cover × Cover) │
└───────────────────────────────┘
```

```
┌─ Pair_set ─────────────────────────────┐
│ F, result_pairs : ℙ (Cover × Cover)    │
│ α₁, α₂ : Cover                          │
└────────────────────────────────────────┘
```

In these schemas, $A_1 \mapsto A_2 \in F$ will be transformed into *res_pairs* as a result of adding a pair coverage element $\alpha_1 \mapsto \alpha_2$. Applying this process to all elements of $F$ gives a new set, called *result_pairs*.

The intersection function is the simplest, and its definition will be given in full. Following the rules for the intersection of Cartesian products, for one element of a pair set the intersection is given by:

```
┌─ Int_element ──────────────────┐
│ Pair_element                    │
│ ┌─────────────────────────────┐ │
│ │ res_pairs = {x ↦ y}         │ │
│ │ where                        │ │
│ │   x == intersection(A₁, α₁) │ │
│ │   y == intersection(A₂, α₂) │ │
│ └─────────────────────────────┘ │
└────────────────────────────────┘
```

The set of pairs is obtained by adding together all the elements and discarding any that are empty:

```
┌─ Int_pair ──────────────────────────────────────────────┐
│  Pair_set                                               │
├─────────────────────────────────────────────────────────┤
│  result_pairs = {Int_element; x, y : Cover              │
│                  | A_1 ↦ A_2 ∈ F                        │
│                  ∧ x ↦ y ∈ res_pairs ∧ x ≠ incomplete ∧ y ≠ incomplete │
│                  • x ↦ y                                │
│                  }                                      │
└─────────────────────────────────────────────────────────┘
```

With this, the definition of *intersection* can be given by cases on the structure of *Cover* as follows:

```
┌─ IPrimitive ────────────────────────────────────────────┐
│  c_1, c_2, c : Cover                                    │
├─────────────────────────────────────────────────────────┤
│  (c_1 = complete ∧ c = c_2) ∨ (c_1 = incomplete ∧ c = incomplete) │
│  ∨                                                      │
│  (c_2 = complete ∧ c = c_1) ∨ (c_2 = incomplete ∧ c = incomplete) │
└─────────────────────────────────────────────────────────┘
```

A complete cover accounts for all values and so the result is unchanged. An incomplete cover corresponds to the empty set which cannot have an intersection with any set of values.

For constructed covers, the intersection is determined by whether the constructors are equal and if so, whether the parameter values intersect:

```
┌─ IConstructions ────────────────────────────────────────┐
│  c_1, c_2, c : Cover                                    │
├─────────────────────────────────────────────────────────┤
│  c_1 ∈ ran construct ∧ c_2 ∈ ran construct              │
│     F' = {} ∧ c = incomplete                            │
│     ∨                                                   │
│     F' ≠ {} ∧ c = construct F'                          │
│  where                                                  │
│     F == construct⁻¹ c_1                                │
│     F' == {c : dom F; parc, int : Cover                 │
│            | c ↦ parc ∈ construct⁻¹ c_2                 │
│            ∧ int = intersection(F c, parc) ≠ incomplete │
│            • c ↦ int                                    │
│            }                                            │
└─────────────────────────────────────────────────────────┘
```

For tuples, any member of the *pair* coverages may intersect:

$$\begin{array}{|l|}
\hline
\textit{ITuples} \\
c_1, c_2, c : \textit{Cover} \\
\hline
c_1 \in \textit{ran pair} \land c_2 \in \textit{ran pair} \\
\quad F' = \{\} \land c = \textit{incomplete} \\
\quad \lor \\
\quad F' \neq \{\} \land c = \textit{pair } F' \\
\textbf{where} \\
\quad F == \textit{pair}^{-1} c_1 \\
\quad F' == \bigcup \{\textit{Int\_pair} \mid \alpha_1 \mapsto \alpha_2 \in \textit{pair}^{-1} c_2 \bullet \textit{result\_pairs}\} \\
\hline
\end{array}$$

Finally, the intersection function is given by the constraint:

$$\begin{aligned}
\textit{intersection} = {}& \lambda\, c_1, c_2 : \textit{Cover} \\
& \bullet\, \mu\, c : \textit{Cover} \\
& \quad \mid \textit{IPrimitive} \lor \textit{IConstructions} \lor \textit{ITuples} \\
& \bullet\, c
\end{aligned}$$

The difference function is similar, except that for tuples, the difference of cartesian product sets is slightly more complicated, and for the case when $c_1$ is complete, but $c_2$ is not, it is necessary to expand $c_1$ into the appropriate representation of completeness. The union function is similar in structure to difference, but this time it is also necessary to take into acount the unique representation of completeness. (The intersection and difference functions cannot generate a *complete* cover, although they may generate an *incomplete* one. For the *union* function, the converse is true.)

# 6 Proof opportunities

The section title is intended to convey the concept of selective proof: proof should be used to increase confidence in those parts of the design which need further investigation, rather than calling for the proof of everything from first principles. In a specification of this nature, there are two sorts of proof opportunity, namely those associated with the consistency of the specification and those associated with the refinement process itself. In the course of developing this implementation, the opportunity has been taken to point out various consistency proofs as the need arises. For example, datatypes should be satisfiable, μ-terms should stand for uniquely existing values and the use of function arrows should be compatible with the axiomatic definition of the functions. The proof requirements for consistency are relatively trivial and do not add greatly to the understanding of the problem. Consequently a proof document should concentrate on the proof opportunities generated by the refinement itself.

Even within this constraint, a sea of theorems emerges as soon as the first few layers of definition are peeled off and it seems inconceivable that the theorems could be generated and proved mechanically. Given this the correct approach seems to be to exhibit the main structure of the proof by writing down a few of the main goals which a proof would entail. For more assurance in the software, more of the proof could be exhibited and some of the goals mechanically proven. However a necessary consequence of this approach would be that some proofs would be carried out on the basis of assumptions both informally made and stated.

As an example of this approach, a few sample goals will be given for the proofs associated with the *coverage* function, for which the main theorem to prove is

> *Pattern_Compatible*
> ⊢
> *Abs_fn(coverage(p), type) = valcover(p, type)*

The theorem contains $p$ and *type* as a parameter, so it has to be shown for all values of these types. The constraint in the hypothesis ensures that the type is determined by the pattern in most cases, so the main proof is by induction over the structure of *Pattern*, with *type* corresponding.

Establishing a theorem of the form $P(p)$, where $p$ is a *Pattern* and P some predicate on $p$, requires the proof of the following base cases:

> ⊢ $P(unitp)$
> $sc : SCONST$ ⊢ $P(sconstp(sc))$
> $v : Variable$ ⊢ $P(var(v))$

and the following induction steps

> *Cons_patt*; $P(patt)$ ⊢ $P(consp(\theta Cons\_patt))$
> $tp : tuple\,Pattern;\ \forall\, p : ran\,tp \bullet P(p)$ ⊢ $P(tupp(tp))$

For the coverage theorem, the base cases are a consequence of the following theorems whose proof is immediate:

> *PCPrimitive*; $c : Cover$; $vals : \mathbb{P}\,Value$; $p = unitp$
> ⊢
> *AIPrimitive* ∧ *CPrimitive* ∧ $vals = \{v : type\ |\ MPrimitive\}$

> *PCPrimitive*; $c : Cover$; $vals : \mathbb{P}\,Value$; $p \in ran\,sconstp$
> ⊢
> *AIPrimitive* ∧ *CPrimitive* ∧ $vals = \{v : type\ |\ MPrimitive\}$

$PCPrimitive; \ c : Cover; \ vals : \mathbb{P} \ Value; \ p \in ran \ var$
$\vdash$
$AIPrimitive \land CPrimitive \land vals = \{v : type \mid MPrimitive\}$

In the theorems above, *PCPrimitive* is a schema defining *Pattern_Compatible* for the case of primitive patterns and values. The schemas beginning *AI*, *C* and $\overline{M}$ define similar cases for the abstraction invariant, the *coverage* function and the *matches* relation. The *valcover* function has been expanded in terms of its definition in terms of *matches*, the proof expanded by cases and the irrelevant parts thinned.

For the induction steps, the coverage property required may be expressed using the following schema:

```
┌─ Coverage_property ──────────────────┐
│ p : Pattern; type : Type             │
├──────────────────────────────────────┤
│ Pattern_Compatible ⇒                 │
│   Abs_fn(coverage(p), type) = valcover(p, type) │
└──────────────────────────────────────┘
```

The constructor case will then follow from the theorem:

$p : Pattern; \ type : Type; \ c : Cover; \ vals : \mathbb{P} \ Value$
$Cons\_patt; \ p = consp \ \theta Cons\_patt$
$PCConstructions; \ partype : Type; \ partype = type\_of\_con \ con$
$Coverage\_property_{[patt/p, \ partype/type]}$
$\vdash$
$AIConstructions \land CConstructions \land vals = \{v : type \mid MConstructions\}$

while the tuple case follows from the theorem:

$p : Pattern; \ type : Type; \ c : Cover; \ vals : \mathbb{P} \ Value$
$tp : tuple \ Pattern; \ p = tupp \ tp$
$PCTuples$
$tt : seq \ Type; \ tt = \{i : dom \ tp \bullet i \mapsto \{tv : tupv^{-1} \ [\![type]\!] \bullet tv \ i\}\}$
$\forall \ i : dom \ tp; \ p : ran \ tp; \ type : ran \ tt$
$\mid p = tp \ i \land type = tt \ i$
$\bullet Coverage\_property$
$\vdash$
$AITuples \land CTuples \land vals = \{v : type \mid MTuples\}$

In both of these theorems, the hypothesis list makes use of the fact that the type in the induction hypothesis must be derived from the given type, which is compatible with the pattern. This is expressed by the following theorems:

$p : Pattern; \ type : Type$
$Cons\_patt; \ p = consp \ \theta Cons\_patt$
$Pattern\_Compatible_{[patt/p, \ partype/type]}$
$\vdash$
$partype = type\_of\_con \ con$

$p$ : Pattern; type : Type
$tp$ : tuple Pattern; $p = tupp\ tp$
$tt$ : seq Type
$\forall\ i$ : dom $tp$; $p$ : ran $tp$; type : ran $tt$
$|\ p = tp\ i \land type = tt\ i$
• Pattern_Compatible
⊢
$tt = \{i : dom\ tp \bullet i \mapsto \{tv : tupv^{-1}\ [\![type]\!] \bullet tv\ i\}\}$

The proof of the constructor case may be carried out along the following lines. The conclusion consists of a conjunction of three predicates, each of which must be shown to be true. The second provides a value for the cover, $c$, which may be used to rewrite the first predicate to give

$p$ : Pattern; type : Type; $c$ : Cover; vals : $\mathbb{P}$ Value
Cons_patt; $p = consp\ \theta Cons\_patt$
partype : Type; partype = type_of_con con
PCConstructions; Coverage_property$_{[patt/p,\ partype/type]}$
⊢
$vals = \{val : Abs\_fn(coverage\ patt,\ type\_of\_con\ con) \bullet consv\ \theta Cons\_val\}$
$vals = \{val : type\_of\_con\ con\ |\ patt\ matches\ val \bullet consv\ \theta Cons\_val\}$

This theorem follows immediately from the hypothesis. The tuple case requires an induction over the length of the tuple, starting with the base case of the tuple size being 2. The constructor and tuple case together give the theorem required.

The other requirements on *coverage* are easily established. As an example, we can try and establish

Pattern_Compatible; $c$ : Cover; vals : $\mathbb{P}$ Value
$c = coverage(p) \land vals = valcover(p,\ type)$
⊢
$c = incomplete \lor vals \neq \{\}$

The base cases are a consequence of the following theorems whose proof is immediate.

PCPrimitive; CPrimitive; vals : $\mathbb{P}$ Value
$p = unitp \land vals = \{v : type\ |\ MPrimitive\}$
⊢
$c \neq incomplete \land vals = \{unitv\}$

PCPrimitive; CPrimitive; vals : $\mathbb{P}$ Value
$p \in ran\ sconstp \land vals = \{v : type\ |\ MPrimitive\}$
⊢
$c = incomplete$

PCPrimitive; CPrimitive; vals : $\mathbb{P}$ Value
$p \in ran\ var \land vals = \{v : type\ |\ MPrimitive\}$
⊢
$c \neq incomplete \land vals = type$

The induction property is

```
┌─ Incomplete_property ─────────────────────────────┐
│ p : Pattern;  type : Type                         │
├───────────────────────────────────────────────────┤
│   Pattern_Compatible ⇒ c ≠ incomplete ⇒ vals ≠ {} │
│ where                                             │
│   c == coverage(p)                                │
│   vals == valcover(p, type)                       │
└───────────────────────────────────────────────────┘
```

The constructor induction step is as follows

$p$ : *Pattern*;  *type* : *Type*;  $c$ : *Cover*;  *vals* : $\mathbb{P}$ *Value*
*Cons_patt*;  $p = consp\ \theta Cons\_patt$
*PCConstructions*;  *partype* : *Type*;  *partype* = *type_of_con con*
*Coverage_property*$_{[patt/p,\ partype/type]}$
⊢
*CConstructions* ∧ *vals* = {$v$ : *type* | *MConstructions*}
$c \neq incomplete \Rightarrow coverage\ patt \neq incomplete$

From the conclusion and the hypothesis it is possible to show that *valcover(patt, type_of_con con)* is not empty, from which it follows that *vals* is not empty. As before, the tuple case will involve an induction over the size of the tuple.

Thus, for the relatively trivial case of the *coverage* function, the proofs of 4 theorems are required, as listed in section 5, of which the first two have been outlined here. The third theorem follows immediately from the definition of *Abs_fn* while the fourth will have a similar structure to that given above. Thus to show the main structure of the proof fully we are required to display 18 theorems: 5 each for the non-trivial theorems corresponding to the constructors of *Pattern*, 2 for the type lemmas and one for the third, trivial, theorem.

The *union* function is considerably more complicated. The main theorem we have to show is

*Union*
⊢
*Abs_fn(c, type)* = *Abs_fn(c$_1$, type)* ∪ *Abs_fn(c$_2$, type)*

Two lemmas are needed for this, which prove that the difference and intersection functions have their intended effect. The intersection lemma is the easiest of these, and may be expressed using the schema below:

```
┌─ Intersection ─────────────────────────────────┐
│ c₁, c₂, c : Cover                              │
│ type : Type                                    │
├────────────────────────────────────────────────┤
│ c₁ cover_compatible type ∧ c₂ cover_compatible type │
│ c = intersection(c₁, c₂)                       │
└────────────────────────────────────────────────┘
```

The theorem required is

*Intersection*
⊢
*Abs_fn(c, type)* = *Abs_fn(c$_1$, type)* ∩ *Abs_fn(c$_2$, type)*

The proof involves a structural induction over a cross-product of *Cover*, again with *type* corresponding. There are 4 constructors in the datatype and consequently 16 cases to prove. Of these, the two cases with mixed *pair* and *construct* constructors may be eliminated because they cannot be simultaneously compatible with the type. Of the 14 remaining cases, 12, which deal with the $c_1$ or $c_2$ being *complete* or *incomplete*, are trivial and follow immediately from the fact that the abstraction function delivers the full set of values in the type, or the empty set, respectively.

Altogether, for the *union* case, five proofs are required, as listed in section 5.2, of which the second and fourth are immediate. The first proof breaks down into the *intersection*, *difference* and *union* cases, each based on a cross product of *Cover* each requiring the proof of 16 subsidiary cases, 48 in all. There will also be 2 subsidiary lemmas for manipulating the type in the induction property, just as for *coverage*. The third and fifth theorems also give rise to 16 cases so there are 82 subsidiary goals to establish. Of these the tuple cases are relatively complicated and would need to be broken down into further goals.

# 7 Conclusions

Before discussing the implications of this case study, it is worth emphasising the fact that this report is about refinement, not about the precise form of the pattern matching algorithm to use in ML. Other studies, for example Baudinet and MacQueen [1987], Peyton Jones [1987], give algorithms for compiling pattern matching expressions, but the problem of demonstrating the correctness of the algorithm still remains. Given this approach to the formal development of software therefore, the questions arise as to whether it is actually helpful to the developers and whether it actually convinces the evaluators.

Taking the developers point of view first of all, when reading this case study, one motors along quite happily through sections 1 to 4 and then the going gets rough in section 5 and finally one gets bogged down in section 6. Section 5 is difficult because the problem is difficult. Although the intuitive idea of an exhaustive check is clear it is extremely hard to think of a general algorithm which convincingly copes with all cases. In addition, intuitively one thinks that the type structure must play a relatively minor role in the algorithm, but equally it must be present in the specification and the problem is to see what assumptions can be made about it. Given that, the difficulty in section 5 is quite understandable and this approach, namely the development of the abstraction invariant and its use in calculating the implementation required, is a sensible way of designing software. Note incidentally, that the "bottom-up" approach of designing the software and *then* attempting to show that it satisfies the specification is dangerous. Our work on this study was impeded by several months fruitlessly trying to prove that our intuitive algorithm was correct when it was not.

Given the investment in abstract design, represented by sections 1 to 5, the implementation becomes very easy. The implementation was written, tested and debugged in 1 day. The main structure of the implementation followed closely that of the specification so that it was easy to see how the two related. However the formal distance, in terms of the number of theorems needed to demonstrate the refinement was large. Further steps of data refinement were undertaken. The constructor coverage measure, which in Z has been represented by $CON \nrightarrow Cover$, was instantiated as $\mathbb{N} \nrightarrow Cover$, and implemented as the Algol68 array of cover, []COVER. In addition, the implementation uses a total function from the set of constructors in the datatype, rather than a partial function whose domain is the set of constructors already encountered. In the implementation, the constructors which have not been encountered are given an *incomplete* coverage. For tuples, the Z representation of *pair* 《 $\mathbb{F}$ *(Cover × Cover)* 》 is implemented by the Algol68 mode PAIR = STRUCT(COVER a, b, REF PAIR next). In this case the abstract sets are being implemented by linked lists, with a natural implementation of universal quantifiers in terms of loops.

For the *union* function, there was some operation refinement of the test for completion, the calculation of the remainder coverage and the test for whether the coverage has changed. These are all done during the execution of the main loop in the function, rather than sequentially as would result from a simple-minded implementation of the specification. As a result the *union* function delivers both the new coverage and an indication of whether it has changed.

Needless to say some trivial faults were found such as writing TRUE instead of FALSE and omitting to move on the list pointers in some of the loops. These faults will not be found by the formal method without *much* more formalism, but are relatively easily found by testing. One can argue therefore that abstract algorithm design is cost effective as a production method, but this does not apply to the formal proof aspects detailed in section 6. This is because the actual development of the proof is demanding. And it is demanding because it is boring and it is boring because it does not lead to a much greater insight into the problem.

From the evaluator's point of view one can take a similar warm view to the process of abstract algorithm design exemplified by the first 5 sections, because these help the evaluator to understand the algorithm. Indeed, it is probably the only approach from which one can derive a convincing argument for correctness and consequently the only approach likely to satisfy an independent evaluation. It is however not clear the extent to which the proof process adds further assurance: at some points it is hard to see what is going on through all the detail of the formalism. Furthermore, it seems inevitable that the development will be informal at some stage, simply by reason of the number of proofs required. Section 6 covers only an outline of the major proof steps so one could anticipate a proof document for what is a small part of an ML compiler being perhaps 10 times larger than this section. Moreover, this section is notable for what is omitted rather than what is included. At the higher level the formalisation of how the compiling functions are called has been omitted and at the lower level, there is at least as much operation refinement required to end up with the implementation language as the data refinement needed to end up with the design. The sheer number of proof steps generated makes it seem inconceivable that the pattern matching algorithm in an ML compiler will *ever* be carried out fully formally and mechanically checked.

This does not mean that the formal method is invalid. At the higher level, not much would be gained by formalising the syntax analyser and relating it to the compiling operation specifications so that the pre-conditions for these operations were derived formally, rather than informally as has been done here. Syntax analysis is normally treated with a special purpose tool anyway and the pre-conditions are relatively easily checked by eye. The further refinement involved in arriving at the Algol68 implementation is more debatable. For the most part this is straightforward although one would like to see some code level verification of the loops in the *union* function. Note that the design specification is suggestive of the annotations which would be necessary for verification condition generation by tools such as MALPAS and Gypsy.

Even with mechanical aids, it is unlikely that realistic problems will be capable of having the proof of all the theorems exhibited being carried out mechanically. For this particular case study, going down to this level of detail gives rise to at least a 100 theorems. Requiring the proof of all these theorems to be carried out mechanically for certification would be unreasonably costly. However a selective approach is sensible and enables the level of assurance to be related to the amount of effort spent in verification. For example, in this case study, the theorems are arranged in a tree: the refinement obligations are expressed by two theorems, which are broken down into 9 which expand into the 100 or so. A satisfactory technique would be to prove that the 9 theorems entailed the refinement obligations, thus demonstrating that all cases had been considered and then to take one of the branches of the proof tree, perhaps for the harder tuple case, down to the leaves of the tree. Even this might be too onerous and an evaluator could be satisfied with proving some, but not all, of the theorems on the way, simply in order to make the most cost-effective use of his time. A mechanical prover could help here by proving certain of the easier cases, so that these could simply be accepted by the evaluator.

To summarise:

- The formal development process, in particular the key step of exhibiting the abstraction invariant and using it to develop the abstract algorithm design, is a useful and cost-effective technique for developing software, quite apart from its use in demonstrating correctness.

- Formal proof adds to the assurance, but complete formality obscures. Proof needs to be undertaken within a context using both formal and informal elements. The structure of the proof and the way it is delivered to an evaluator (the proof document) ought to follow intuitive, informal, proof methods.

# References

Baudinet M and MacQueen D (1987). Tree pattern matching for ML, in *Functional Programming Languages and Computer Architectures*, Gilles Kahn (ed), Lecture Notes in Computer Science 274, Springer Verlag.

Gordon M J C, Milner R and Wadsworth C P (1979). Edinburgh LCF, Lecture Notes in Computer Science 78, Springer Verlag.

Harper R, Milner R and Tofte M (1988). The definition of standard ML version 2. Laboratory for the Foundations of Computer Science, report ECS-LFCS-88-62, University of Edinburgh.

Hayes I, (1987). Specification case studies, Prentice Hall International series in Computer Science, 1987.

Jones, C B (1986). Systematic software development using VDM, Prentice-Hall.

King S, Sorensen I H, Woodcock J, (1987). Z: grammar and concrete and abstract syntaxes, Programming Research Group, University of Oxford.

Macdonald R, Randell G P and Sennett C T (1989). Pattern matching in ML - a case study in refinement. RSRE report 89004.

Morgan C C, Robinson K A (1987). Specification statements and refinement, IBM Journal of Research and Development, 31, 5.

Morgan C C (1988). The specification statement, TOPLAS 10, 3.

Paulson L C (1987). Logic and computation. Cambridge tracts in theoretical computer science 2, Cambridge University Press.

Peyton Jones S L (1987). The implementation of functional programming languages, Prentice Hall International series in Computer Science.

Sennett, C T (1987). Review of the type checking and scope rules of the specification language Z, RSRE report 87017.

Spivey, J M (1988). Understanding Z: a specification language and its formal semantics, Cambridge University Press.

Sufrin, B (1983). Formal system specification - notation and examples, in *Tools and Notations for Program Construction* (Neel ed.), Cambridge University Press.

# Author Index

Ahmed, S. N. .................................................................................. 73
Back, R. J. R. ................................................................................... 9
Cheng, J. H. .................................................................................. 51
George, C. W. .............................................................................. 155
Hoare, C. A. R. .............................................................................. 33
Jones, C. B. ................................................................................... 51
Jones, G. ..................................................................................... 133
Milne, R. E. ................................................................................. 155
Morris, J. M. .................................................................................. 73
Sannella, D. .................................................................................. 99
Sennett, C. T. .............................................................................. 171
Sere, K. ........................................................................................... 9
Sheeran, M. ................................................................................. 133